T0340105

In Vitro Toxicology

In Vitro Toxicology

Edited by

Alok Dhawan, PhD
Director, CSIR-Indian Institute of Toxicology Research
Lucknow, India

and

Seok Kwon, PhD
Principal Scientist, Central Product Safety
Global Product Stewardship
Singapore Innovation Center
Procter & Gamble International Operations
Singapore

ACADEMIC PRESS

An imprint of Elsevier

Academic Press is an imprint of Elsevier
125 London Wall, London EC2Y 5AS, United Kingdom
525 B Street, Suite 1800, San Diego, CA 92101-4495, United States
50 Hampshire Street, 5th Floor, Cambridge, MA 02139, United States
The Boulevard, Langford Lane, Kidlington, Oxford OX5 1GB, United Kingdom

Notices
Knowledge and best practice in this field are constantly changing. As new research and experience
broaden our understanding, changes in research methods, professional practices, or medical
treatment may become necessary.

Practitioners and researchers must always rely on their own experience and knowledge in
evaluating and using any information, methods, compounds, or experiments described herein. In
using such information or methods they should be mindful of their own safety and the safety of
others, including parties for whom they have a professional responsibility.

To the fullest extent of the law, neither the Publisher nor the authors, contributors, or editors,
assume any liability for any injury and/or damage to persons or property as a matter of products
liability, negligence or otherwise, or from any use or operation of any methods, products,
instructions, or ideas contained in the material herein.

Library of Congress Cataloging-in-Publication Data
A catalog record for this book is available from the Library of Congress

British Library Cataloguing-in-Publication Data
A catalogue record for this book is available from the British Library

ISBN: 978-0-12-804667-8

For information on all Academic Press publications visit our website at
https://www.elsevier.com/books-and-journals

Working together
to grow libraries in
developing countries

www.elsevier.com • www.bookaid.org

Publisher: Mica Haley
Acquisition Editor: Rob Sykes
Editorial Project Manager: Tracy Tufaga
Production Project Manager: Mohanambal Natarajan
Designer: Mark Rogers

Typeset by TNQ Books and Journals

Contents

List of Contributors   xi
Editors Biographies   xiii
Foreword   xv
Preface   xvii
Acknowledgments   xix

1. Development of In Vitro Toxicology: A Historic Story

Shripriya Singh, Vinay K. Khanna and Aditya B. Pant

Introduction   1
What In Vitro Toxicology and How It Came Into Being?   1
Why In Vitro Not In Vivo?   2
The Regulatory Control and Policy Makers   2
Projects and Developments   4
Some of the Validated In Vitro Alternatives and Methods   5
The Available In Vitro Models Systems   7
Explant/Organ Culture   7
Cell Culture   8
Primary Cell Culture   8
Immortalized Cell Lines   8
Organotypic Cultures   8
Stem Cells and Induced Pluripotent Stem Cells   9
Organoid Culture   9
Basal to Organ-Specific Endpoints of In Vitro Toxicology   9
Basal Toxicity or Cytotoxicity   9
Genotoxicity   10
Cells to 3D Organoids: In Vitro Organ System Toxicology   11
Hepatoxicity In Vitro   11
Neurotoxicity In Vitro   12
Nephrotoxicity In Vitro   13
Cardiotoxicity In Vitro   14
Nanotoxicology In Vitro   14
Summary   15
Suggested Web Portals for Further Reading   15
References   16

2. **Principles for In Vitro Toxicology**

Sonal Srivastava, Sakshi Mishra, Jayant Dewangan, Aman Divakar, Prabhash K. Pandey and Srikanta K. Rath

Introduction	21
Guidelines Governing the In Vitro Methods for Toxicity Testing	22
Optimization of Cell Culture	25
Models of In Vitro Toxicity	26
Explant Culture	27
Organotypic Culture	27
Cell Cultures	27
Three Dimensional Culture Systems	28
Development of Biomarkers for In Vitro Toxicity Evaluation	28
Assays for Screening of Toxicants	29
Ames Test	29
The Hen's Egg Test–Chorio Allantoic Membrane Test	29
Corrosivity Assays	32
3T3 Neutral Red Uptake Test	32
Tetrazolium Salt, 3-(4,5-Dimethylthiazol-2-yl)-2, 5-Diphenyltetrazolium Bromide Assay	32
Alamar Blue	33
The Bovine Cornea Opacity/Permeability Test	33
Dose–Response Relationship	33
Evaluation of DR Relationship	34
Data Interpretation	35
Validation of In Vitro Toxicology Test Methods	36
The "Omics" Approach	37
Conclusion	38
Acknowledgments	38
References	40

3. **Models and Methods for In Vitro Toxicity**

Abhishek K. Jain, Divya Singh, Kavita Dubey, Renuka Maurya, Sandeep Mittal and Alok K. Pandey

Introduction	45
Need of In Vitro Models for Toxicity Assessment	46
Validation of In Vitro Test Methods	47
Criteria for Validation	47
Criteria for Regulatory Acceptance	47
Models for In Vitro Toxicity Assessment	48
In Vitro Models for Cytotoxicity Studies	48
In Vitro Models for Specific Toxicity	48
Methods Employed for In Vitro Toxicity Assessment	51
Cytotoxicity Assessment	51
Genotoxicity Assessment	53
Mutagenicity	56
Toxicokinetic Study	58

Cellular and Functional Responses (Protein/Gene Expression)   58
State of Art and Recent Development Regarding In Vitro
Approaches   59
Omics Approach   59
Bioinformatics and Computational Toxicology   60
Integrated Testing Strategies   60
Challenges and Considerations   60
Future Perspective   61
Conclusion   62
References   62

4. In Vitro Gene Genotoxicity Test Methods

Manisha Dixit and Amit Kumar

Introduction   67
Importance of Genotoxicity Testing   68
Test for Genotoxicity   70
The Standard Test Battery for Genotoxicity   70
General Protocol for In Vitro Tests   72
Different In Vitro Tests   74
Bacterial Reverse Mutation Test (TG 471)   74
In Vitro Mammalian Chromosome Aberration Test (TG 473)   76
In Vitro Mammalian Cell Gene Mutation Tests Using the
  Hprt and *xprt* Genes (TG 476)   77
In Vitro Mammalian Cell Micronucleus Test (TG 487)   79
In Vitro Mammalian Cell Gene Mutation Tests Using the
  TK Gene (TG 490)   80
Current Limitations of In Vitro Genotoxicity Testing   82
Faulty Metabolic Activator System   82
False Positive Results   83
False Negative Results   83
Future Prospective of the Genotoxicity Testing   84
Summary   84
Acknowledgments   85
References   85

5. In Silico Approaches for Predictive Toxicology

Ramakrishnan Parthasarathi and Alok Dhawan

Introduction   91
In Silico Toxicology Framework for Toxicity/Safety Prediction   91
In Silico Toxicology   93
QSAR for Predictive Toxicology   95
Descriptors for Predictive Toxicology   97
Databases and Web Tools for Predictive Toxicology   100
Summary   106
Acknowledgments   106
References   107

6.   **The Use of Transcriptional Profiling in In Vitro Systems to Determine the Potential Estrogenic Activity of Chemicals of Interest**
   Jorge M. Naciff

Introduction	111
Why Do We Have to Assess the Potential Estrogenicity of Chemicals of Interest?	112
Transcriptional Profiling in In Vitro Systems to Determine the Potential Estrogenic Activity of Chemicals of Interest	115
Connectivity Mapping to Define Mode of Action	120
In vitro to in Vivo Extrapolation	121
Conclusion	122
References	123

7.   **Extrapolation of In Vitro Results to Predict Human Toxicity**
   Sonali Das

Need to Understand Toxicity	127
Organ Level and System Level Toxicity	128
Hepatotoxicity	128
Toxicity Assessment In Vitro: Advantages and Disadvantages	131
In Vitro Systems for Liver Toxicity Assessment	131
In Vitro System for Nephrotoxicity Assessment	136
To Overcome In Vitro Assay Limitation	136
Extrapolation of In Vitro Information to In Vivo for Human: Successful and Unsuccessful Outcomes	138
Upcoming Potential Methods to Predict Toxicity	138
References	139

8.   **Role of Molecular Chaperone Network in Understanding In Vitro Proteotoxicity**
   Tulika Srivastava, Sandeep K. Sharma and Smriti Priya

Introduction	143
Balance Between Protein Folding and Misfolding: Misfolded Proteins Are Toxic Species	144
Mechanisms of Cellular Proteotoxicity	145
Cellular Defense Against Proteotoxicity: Molecular Chaperones	146
Role in Proteotoxicity: Molecular Chaperones Antagonizes the Accumulation of Proteotoxic Proteins	147
The Cellular Degradation Mechanism in Proteotoxicity	151
The Ubiquitin Proteasome System	151
Autophagy	153
Specific Role of Degradation Pathways in Mitigating Proteotoxicity	154
Toxicity of Amyloid Beta in AD	155

Toxicity of α-Synuclein in PD   155
Toxicity of Huntingtin Protein in HD   156
Prion Proteins in Encephalopathies   157
Conclusion   158
References   159

9. Scientific and Regulatory Considerations in the Development of in Vitro Techniques for Toxicology

Mukul R. Jain, Debdutta Bandyopadhyay and Rajesh Sundar

Introduction   165
Why In Vitro Toxicity?   165
In Vitro Assays for Safety and Toxicity Assessment   166
Cytotoxicity Assessment   166
Assessment of Carcinogenicity   169
Reproductive and Developmental Toxicity Assessment   171
Phototoxicity Assessment   172
Hepatotoxicity   172
Safety Pharmacology   173
Tissue Cross-Reactivity Assay   173
The Application of the Principles of GLP to in Vitro Studies   174
Validation of in Vitro Methods   176
Regulatory Acceptance of Alternative Methods to Animal Testing 178
Translational Aspects of in Vitro Assays   178
Pitfalls in In Vitro Toxicology   178
Acknowledgments   183
References   183

10. Safety Concerns Using Cell-Based In Vitro Methods for Toxicity Assessment

Vyas Shingatgeri

Introduction   187
Biosafety and Risk Management in In Vitro Laboratories   188
Risk-Classification Based on Type of Cells   189
Risk Associated With Cell Culture   189
Who May Be at Risk?   190
Factors Contributing Toward Risk in the Cell Culture   190
Tissue/Type, Origin, and Source of Cells   190
Viral Contaminanation   191
Problems Associated With Viral Contaminations   192
Origin of Viral Contaminations   193
Mycoplasmas   193
How Does Mycoplasma Enter Cell Cultures?   194
Culture Media   194
Transmission Involving Use of Needles and Sharp Objects
    During Manipulations   195

Fetal Bovine Serum 195
Handling Cells Outside Biosafety Cabinets 195
Altering Culture Conditions 196
Risk Associated With Procedures Involving Cryobanks/
   Liquid Nitrogen 196
Risk Associated With Procedures Involving CO_2 Cylinders 196
Risk Associated With Procedures Involving Toxic Compounds 197
Physical Hazards and Others 198
Ergonomic Hazards 198
Ionizing Radiations 198
Nonionizing Radiations 198
Noise Hazards 198
Recommended Practices 199
Containment Measures for Working With Cell Cultures 201
State and Federal Regulations 201
OSHA Standards 202
Laboratory Standards 202
Hazard Communication Standards 202
The Blood Borne Pathogens Standard 202
The Personal Protective Equipment Standard (29 CFR 1910.132) 202
The Eye and Face Protection Standard (29 CFR 1910.133) 203
The Respiratory Protection Standard (29 CFR 1910.134) 203
The Hand Protection Standard (29 CFR 1910.138) 203
The Control of Hazardous Energy Standard (29 CFR 1910.47) 203
Conclusions 203
References 205
Further Reading 207

11. In Vitro Methods for Predicting Ocular Irritation

Sanae Takeuchi and Seok Kwon

Introduction 209
In Vitro Methods Adopted by OECD 210
Isolated Chicken Eye Test 210
Bovine Corneal Opacity and Permeability Test 211
Fluorescein Leakage Assay 212
Short Time Exposure Test 213
Reconstructed Human Cornea-Like Epithelium Test Method 213
Recent Development and Looking Into Future 214
Conclusion 216
References 216

Index 221

List of Contributors

Debdutta Bandyopadhyay Zydus Research Centre, Ahmedabad, India

Sonali Das Syngene International Pvt. Ltd., Bangalore, India

Jayant Dewangan CSIR-Central Drug Research Institute, Lucknow, India

Alok Dhawan CSIR-Indian Institute of Toxicology Research, Lucknow, India

Aman Divakar CSIR-Central Drug Research Institute, Lucknow, India

Manisha Dixit CSIR-Indian Institute of Toxicology Research, Lucknow, India

Kavita Dubey CSIR-Indian Institute of Toxicology Research, Lucknow, India

Abhishek K. Jain CSIR-Indian Institute of Toxicology Research, Lucknow, India

Mukul R. Jain Zydus Research Centre, Ahmedabad, India

Vinay K. Khanna CSIR-Indian Institute of Toxicology Research, Lucknow, India

Amit Kumar CSIR-Indian Institute of Toxicology Research, Lucknow, India

Seok Kwon Procter & Gamble International Operations SA Singapore Branch, Singapore

Renuka Maurya CSIR-Indian Institute of Toxicology Research, Lucknow, India

Sakshi Mishra CSIR-Central Drug Research Institute, Lucknow, India

Sandeep Mittal CSIR-Indian Institute of Toxicology Research, Lucknow, India

Jorge M. Naciff The Procter and Gamble Company, Mason, OH, United States

Alok K. Pandey CSIR-Indian Institute of Toxicology Research, Lucknow, India

Prabhash K. Pandey CSIR-Central Drug Research Institute, Lucknow, India

Aditya B. Pant CSIR-Indian Institute of Toxicology Research, Lucknow, India

Ramakrishnan Parthasarathi CSIR-Indian Institute of Toxicology Research, Lucknow, India

Smriti Priya CSIR-Indian Institute of Toxicology Research, Lucknow, India

Srikanta K. Rath CSIR-Central Drug Research Institute, Lucknow, India

Sandeep K. Sharma CSIR-Indian Institute of Toxicology Research, Lucknow, India

Vyas Shingatgeri Vanta Bioscience Limited, Chennai, India

Divya Singh CSIR-Indian Institute of Toxicology Research, Lucknow, India

Shripriya Singh CSIR-Indian Institute of Toxicology Research, Lucknow, India

Sonal Srivastava CSIR-Central Drug Research Institute, Lucknow, India

Tulika Srivastava CSIR-Indian Institute of Toxicology Research, Lucknow, India

Rajesh Sundar Zydus Research Centre, Ahmedabad, India

Sanae Takeuchi P&G Innovation Godo Kaisha, Kobe, Japan

Editors Biographies

Alok Dhawan
CSIR-Indian Institute of Toxicology Research, Lucknow, India

Professor Alok Dhawan is currently Director, CSIR-IITR (Council of Scientific & Industrial Research-Indian Institute of Toxicology Research), Lucknow and Outstanding Professor, Academy of Scientific and Innovative Research, New Delhi. He is the test facility management of the Good Laboratory Practice compliant toxicity test facility at CSIR-IITR. He served as the founding Director, Institute of Life Sciences, and dean, Planning and Development, Ahmedabad University, Gujarat. Before joining as director, CSIR-IITR, he worked as a scientist C, principal scientist, and senior principal scientist at CSIR-IITR, Lucknow. He obtained his Ph.D. degree in biochemistry from the University of Lucknow, India, in 1991 and was awarded D.Sc. degree (*h.c.*) from University of Bradford, UK, in 2017. He has been a visiting scholar, Michigan State University, USA; Boycast fellow, University of Surrey, Wales and Bradford, UK. He has over 30 years of research experience in toxicology and has contributed significantly in the areas of in vitro, genetic, and in silico toxicology. Professor Dhawan initiated the area of nanomaterial toxicology in India and published a guidance document on the safe use of nanomaterials. His group elucidated the mechanism of toxicity of metal oxide nanoparticles in human and bacterial cells and also looked at the environmental impact of nanomaterials.

Professor Dhawan has won several honours and awards including the INSA (Indian National Science Academy) Young Scientist Medal in 1994, CSIR Young Scientist Award in 1999, the Shakuntala Amir Chand Prize of ICMR (Indian Council of Medical Research) in 2002, and the Vigyan Ratna by the Council of Science and Technology, UP in 2011. His work in the area of nanomaterial toxicology has won him international accolades as well and he was awarded two Indo-UK projects under the prestigious UK-IERI (India Education and Research Initiative) program. He was also awarded two European Union Projects under the FP7 and New INDIGO programs.

He founded the Indian Nanoscience Society in 2007. In recognition of his work, he has been elected Fellow, Royal Society of Chemistry, UK; Fellow, The National Academy of Sciences, India; Fellow, The Academy of Toxicological Sciences, USA; Fellow, The Academy of Environmental Biology; Fellow, Academy of Science for Animal Welfare; Fellow—Society of Toxicology (India); founder Fellow, Indian Nanoscience Society; Fellow,

Gujarat Science Academy; elected Fellow, National Academy of Medical Sciences, India; president, Uttar Pradesh Academy of Sciences (2017-); Vice President—Environmental Mutagen Society of India (2006–2007); member, United Kingdom Environmental Mutagen Society, United Kingdom; member, Asian Association of Environmental Mutagen Societies, Japan.

He has to his credit over 130 publications in peer-reviewed international journals, 18 reviews/book chapters, four patents, two copyrights, and has edited four books. He serves on the editorial boards of several scientific journals of repute such as Mutagenesis, Nanotoxicology, Xenobiotica, and Mutation Research.

Seok Kwon
Procter & Gamble International Operations, Singapore

Dr. Seok Kwon has been working for Procter & Gamble (P&G) for almost two decades as a human safety toxicologist in Central Product Safety, Global Product Stewardship, Research & Development. He is currently located at Singapore Innovation Center, Procter & Gamble International Operations in Singapore. Dr. Kwon has worked across multiple product categories, including Skin, Hair, Perfumery, Fabric & Home, Paper, Baby and Hair Care for global markets. Dr. Kwon is the Chairman of the Annual Asia Safety Expert Symposium as a forum for the risk communications with external stakeholders. He is currently active in the Safety Advisory Committee for ASEAN Cosmetic Association.

Prior to joining P&G, Dr. Kwon has completed his postdoctoral fellowship in Endocrine, Reproductive & Developmental Toxicology Program at Chemical Industry Institute of Toxicology, North Carolina, USA; a visiting fellowship in Laboratory of Reproductive & Developmental Toxicology at National Institute of Environmental Health Sciences, National Institutes of Health, North Carolina, USA; and PhD from University of Illinois at Urbana–Champaign, USA.

Foreword

"In vitro toxicology" is a timely and important topic in the field of chemical safety assessment. The final goal of toxicology is to assess risk for human of chemicals introduced newly into the market or chemicals existing in our environment for many years. We have used assay systems using experimental animals for chemical safety assessment as, so-called, golden standard. We can get a lot of important information about toxicity of chemicals but we have to extrapolate from the results of animal experiment to human. Because of this limitation, we try to find the profile of toxicity including mechanistic consideration. Moreover, the movement of consideration that animal welfare becomes unavoidable and we should take care of 3R's (Replacement, Reduction, Refinement) spirit and should be included in the animal experiments. To overcome these problems, we are approaching in in silico or computational toxicology, which may be a final goal. However, we do not have any established such tools in our hands and still many approaches are challenging. The approach of in vitro toxicology can be a successful key to fill the gap between animal experiments and computational assessment. The development of in vitro methods for toxicology has improved a lot in the last years. The outcomes from in vitro assays can endorse the results of animal experiments and give important pieces to establish an adverse outcome pathway (AOP). AOP will give the golden base for the risk assessment of chemicals for our lives. This book covers historical review of in vitro toxicology, specific principles, and methods of toxicology and also the future directions. I believe that each chapter will give readers very important information why in vitro approach is essential to complete risk assessment of chemicals for human safety life.

Makoto Hayashi, D.Sc.
Representative
Makoto International Consulting
Ebina, Japan

Preface

A rise in the number of cases related with health anomalies due to environmental toxicants over the years has been a cause of global concern. Consequently, there has been a need to identify persisting chemicals/compounds and their safety levels for devising effective mitigation strategies in the exposed population. Over the years, this has led to development and validation of reliable and predictive methods chemical toxicity as well as risk assessment.

Since its inception, the whole idea of toxicity testing largely focused on laboratory animal models. These model systems represented a complete interconnected organ system so as to clearly study the effects induced by a certain chemical/compound inside the body. However, to reduce the animal testing and improve the predictability of chemical action, in silico and in vitro methods for chemical safety assessment evolved.

However, the scientific validity of each and every model needs to be determined to carry it forward. As a result, guidelines have been laid down by several regulatory agencies to scientifically evaluate and validate alternate models/methods.

This book is a compilation of 11 informative chapters that cover the nuances of in vitro toxicity methods in biomedical and pharmaceutical sciences. The first chapter "Development of in vitro toxicology: A historic story" sheds light on the historical background of in vitro toxicology since the first cell culture in the 19th century and its advancement thereafter. The next two chapters "Principles for in vitro toxicology" and "Models and methods for in vitro toxicity" discuss the principles and methods underlying in vitro toxicity assessment focusing on the need and requirements for establishing in vitro models and how can they be designed effectively. Another chapter "In vitro gene genotoxicity test methods" gives an insight into the genotoxicity assessment methods based on in vitro model system to estimate the DNA damage induced in the exposed population. The other chapters "Scientific and regulatory considerations in the development of in vitro techniques for toxicology", "Safety issues in the use tissues/in vitro methods for toxicity assessment" address the regulatory and safety issues behind the use of in vitro models. "Extrapolation of in vitro results to predict human toxicity" describes about the authenticity of data obtained so that it could be extrapolated for human studies. One of the chapters "In silico approaches for predictive toxicology" covers the in silico approaches as an alternative to animal testing. Other chapters like "In vitro methods for predicting ocular irritation" focus on ocular irritation models developed, and "Role of molecular chaperone network in understanding in vitro proteotoxicity" sheds light over the involvement of chaperones in cases of proteotoxicity through in vitro models. "The use of transcriptional profiling in in vitro systems to determine the potential estrogenic activity of chemicals of interest" is a chapter dealing with chemical-induced alterations or induction of estrogenic activity using in vitro model systems.

This book overall is a rich source of information regarding in vitro toxicology assessment methods to determine chemical safety and the issues related with it. With an impressive lineup of chapters, each being a reservoir of information, this book caters to a broad yet vital target audience. We are sure that it will be an important resource for students and budding young research scholars. It will serve as a guide for the scientists involved in the field of toxicology and a reference book for regulatory affairs people and agencies.

The book is a culmination of several years of friendship between the editors and their long-term association with the authors.

Alok Dhawan
Seok Kwon

Acknowledgments

The editors wish to acknowledge with thanks the hard work and contributions of all the authors in the book. AD would like to acknowledge the generous funding from the Council of Scientific and Industrial Research, India, under its network projects NWP 34, NWP 35, INDEPTH (BSC 0111), and NanoSHE (BSC 0112).

Chapter 1

Development of In Vitro Toxicology: A Historic Story

Shripriya Singh, Vinay K. Khanna, Aditya B. Pant

CSIR-Indian Institute of Toxicology Research, Lucknow, India

INTRODUCTION

"All substances are poisons; there is none which is not a poison. The right dose differentiates a poison from a remedy." This famous quote by Philippus Aureolus Theophrastus Bombastus von Hohenheim Paracelsus, known as the "Father of Modern Toxicology" laid the foundation of this science, which finds widespread application even five centuries later. The biosafety assessment and toxicity profiling of each and every chemical entity present on the earth is required in order to understand this complex science, which is intricately interwoven with the other disciplines such as biology, chemistry, pharmacology, and biomathematics.

Toxicology has been broadly defined as the science of poisons and poisons are often defined as substances that cause harmful effects on administration by either design or accident to living organisms [1]. However, the word "poison" seems a misnomer in the present day as toxicology does not entail a detailed study of only harmful substances but comprises a comprehensive understanding of all drugs, chemical substances, environmental pollutants, pesticides, endocrine disruptors, xenobiotics, and even plant exudates and extracts. Toxicology is an extended arm of medicine and an in-depth understanding of the science is crucial for both core researchers and medical practitioners worldwide.

WHAT IN VITRO TOXICOLOGY AND HOW IT CAME INTO BEING?

In vitro toxicology can be defined broadly as the toxicological phenomena studied in nonwhole animal models. This broad connotation includes tissue slices, isolated organs, isolated primary cell cultures, explants cultures, cell lines, and even subcellular fractions like that of mitochondria, microsomes, and even membranes [2]. Biosafety assessment and toxicity testing is a multitier

In Vitro Toxicology. http://dx.doi.org/10.1016/B978-0-12-804667-8.00001-8

process ranging from in silico, in vitro, in vivo, and finally to the clinical field trials. However, the last decade has seen an upsurge in the in silico and in vitro aspects. The relevance and popularity of the in vitro methods have increased due to several reasons. A lot can be saved both in terms of time and money with in vitro technology. Biotechnological advances have provided us with rapid, reproducible, and reliable assays. The experimental setup is comparatively easy than in vivo and can be managed with respect to time and number of samples. Animal welfare, societal concerns, and ethics in research have been largely pivotal in this paradigm shift from in vivo to in vitro.

WHY IN VITRO NOT IN VIVO?

"Life eats life", a phrase associated with the scriptures, finds relevance even in science, research, and medicine. Humans have depended on animals since times immemorial for their various needs and so have researchers for their simple experiments. Historical records account for animal testing, which goes back to almost centuries when Greeks such as Aristotle and Erasistratus performed experiments on animals in the third and fourth century BC. Ancient Indian Vedic texts such as the "*Sushruta Samhita*," which is considered the foundation of Ayurveda also mentions the use of animal-based medical formulations and as models of dissection and surgery. Drastic technological and scientific advancements have led to increased knowledge about animal breeding and experimentation commonly referred to as the in vivo research. Since then, animal models have become an integral part for testing and assuring the quality and efficacy of pharmaceuticals, vaccines, and biological products and have given us an in-depth understanding of the physiology, anatomy, molecular, and cellular biology. Euthanizing animals for the sake of human welfare is inevitable if scientific and medical advances are to be made. However, indiscriminate animal sacrifice in the name of research seems not only unreasonable, unacceptable but also a moral burden. This has led to the development of animal ethics and adoption of guidelines and animal welfare policies by the entire scientific and medical fraternity. Regulatory bodies are crucially addressing each and every animal-based experimental issue to bring about reforms or to provide for practical and feasible alternatives wherever possible. The world has joint hands to bridge this gap and strike a balance between animal ethics and scientific/medical progress.

THE REGULATORY CONTROL AND POLICY MAKERS

The 3 Rs principles described by WMS Russel and RL Burch way back in 1959 in their book "The Principles of Humane Experimental Technique" were the pioneer guiding principles for animal welfare and ethics in research [3]. The 3 Rs corresponding to Replacement, Reduction, and Refinement clearly describe the use of alternative strategies, reduction in the number of experimental animals, and refinement of the prevalent methods to minimize the suffering and pain of experimental animals. The 3 Rs principles have been universally and unanimously adopted worldwide and have been assimilated into the legislative policies as well as made part of the animal welfare guidelines [4]. As an addition to the existing Rs another R for "Rehabilitation" was brought in picture by CPCSEA (The Committee for the Purpose of Control and Supervision of Experiments on Animals) in India. The fourth R clearly reflects the moral responsibility of researchers toward animals after experimentation. The cost for rehabilitation and postcare of experimental animals should be accounted for while budgets for a research project are made [5]. The need to curb unnecessary animal sacrifice for research has been addressed nearly three decades ago and since then considerable progress has been made to deal with the issue.

The European Centre for the Validation of Alternative Methods (ECVAM) was established in 1991 in Italy and became completely operational in 1993 [6]. The European Union was among the first to propose a complete ban on the use of animals in the testing of cosmetics and since then a number of centers have been established globally in an attempt to develop alternatives to laboratory animals. In India, CPCSEA under the Ministry of Environment and Forests, Government of India, is the authorized body responsible for the screening and approval of the working protocols and methods on animal experimentation.

The approval and validation of the alternative strategies is carried out by three major organizations known as the validation authorities. These are ICCVAM (Interagency Coordinating Committee on the Validation of Alternative Methods), ECVAM, and OECD (Organization for Economic Cooperation and Development).

Some basic steps considered important for validation are briefly discussed,

- **Research and development** is generally funded by government agencies or a regulatory body and rarely by the Industries.
- **Prevalidation** requires almost about 2–3 years and the aim comprises, establishment of the mechanism of a test; optimization and standardization of protocols; evaluation of the intralaboratory variation and prediction of toxicological endpoints via prediction models and data interpretation.
- **Validation** aims for the evaluation of transferability of a test to another laboratory. The process generally takes about 1 year and depends on the availability of an existing peer-review body such as the ECVAM Scientific Advisory Committee, or ESAC, and if not a new ad hoc panel of experts is convened.
- Regulatory acceptances are governed by the law and therefore differ from country to country.
- In Europe, endorsement by ESAC leads to European Union-wide acceptance under applicable rules and regulations, given by the legal requirements under Directive 86/609/EEC that nonanimal alternative strategies should be used preferentially. In the United States, ICCVAM is responsible for formulating recommendations as per the peer review findings and in consultation with

the regulatory agencies and the general public. As per law, ICCVAM's recommendations are to be responded within the next 6 months. However, the process might take 2 or more years at the national level and even longer in the case of internationally consenting bodies such as OECD, International Conference of Harmonization (ICH), and International Cooperation on Harmonisation of Technical Requirements for Registration of Veterinary Medicinal Products (VICH). Further, adherence to the international guidelines is mandatory for acceptance of the data at the international platform.

The Mutual Acceptance of Data agreement is an initiative of the OECD and has led to the harmonization and uniformity of chemical control globally. Reinvention of the wheel has become a common practice, different research groups in different parts of the world carry out identical toxicology studies either deliberately or unintentionally. This often results in repetitive testing and leads to unnecessary animal deaths. According to the MAD agreement the OECD member countries have joint hands to carry out the testing of chemicals uniformly in adherence with the OECD test guidelines and principles of good laboratory practice. Standard operating procedures are followed and documented in a stringent manner so that the quality of data generated is uniform and unanomalous across all laboratories of the world. The data generated in this way shall be accepted by all the OECD member countries and this shall greatly reduce the number of animal deaths per annum. In vitro strategies are widely adopted but regulatory acceptance and validation is mandatory in order for an in vitro alternative to completely replace the existing animal-based studies [7]. However, formal validation is not required if integrated in vitro strategies is used for simple hazard classification, labeling, risk assessment, and screening purposes [8].

ECVAM has been instrumental in bringing about complete replacement methods for many toxicological parameters studied such as skin corrosion, skin irritation, phototoxicity, and skin penetration/absorption. Partial replacement via in vitro methods has also been possible in areas such as that of carcinogenicity, genotoxicity, reproductive toxicity, eye corrosion, eye irritation, and a few more [5]. Many of these validated methods have now been accepted internationally to form a part of the test guidelines of the European Pharmacopoeia and the OECD [8].

PROJECTS AND DEVELOPMENTS

The Seventh Amendment of Council Directive 76/768/EEC based on the approximation of the laws of the Member States clearly defined a step by step phase out of the animal tested cosmetic products over a span of 10 years. This was followed by a complete marketing ban on the cosmetic products tested on animals, which was enforced within the European Union in March 2013. Subsequently, on May 24, 2014, the regulatory body in India, i.e., Drug Controller, Government of

India, has also posed a ban on use of animals for toxicity evaluation of ingredients to be used in cosmetics. This eventually led to the development of a joint research initiative: safety evaluation ultimately replacing animal testing-1 between the Cosmetics Europe and the European Commission, which aims for the development of alternate in silico and in vitro approaches for the repeated dose systemic toxicity testing conducted on animals employed in the assessment of human safety. The initiative comprises six independent research projects related to themes such as organ-simulating devices (HeMiBio), computational modeling (COSMOS), systems biology (NOTOX), human-relevant biomarker detection (DETECTIVE), human stem cell technology (SCR and TOX), integrated data analysis (ToxBank), and COACH, which is an umbrella coordination and support action project. The overall aim was to develop novel and sophisticated in silico and in vitro approaches as successful alternatives [9]. The project DETECTIVE is aimed at investigating human biomarkers and endpoints for repeated dose toxicity via the in vitro cellular models. High-throughput screening platforms and "-omics" technology has been employed to generate data in the alternatives, which shall be in relevance with that of humans.

Another global joint venture was the Sens-it-iv project comprising 28 research groups from industry, research, and academia. Spanning over a period of 66 months the project dealt with the development of alternative in vitro strategies for the assessment and screening of respiratory and skin sensitization by allergens mostly proteins and chemicals. The project was successfully accomplished in 2011 and has thus yielded several advanced alternative methods for which standard operating procedures have been documented [10].

SOME OF THE VALIDATED IN VITRO ALTERNATIVES AND METHODS

Genotoxicity testing was probably one of the first few areas of toxicology where applicability of in vitro strategies received acceptance. In vitro phototoxicity assay was the first cellular assay which got acceptance apart from that for mutagenicity in 2000 by the European Commission. This assay employed 3T3, a mouse fibroblast cell line and used neutral red uptake as a cytotoxicity end point. Traditional toxicity tests such as the median lethal dose (LD50) have been banned and have been replaced with the in vitro techniques. LD50 corresponds to the chemical concentration, which causes 50% death of the experimental animals. IC50 is an in vitro test similar to LD50, the test measures cytotoxicity and corresponds to a chemical's concentration, which inhibits 50% of cell population.

The Draize test for eye irritation in rabbits faced criticism for the cruelty inflicted on test animals and has been universally replaced by in vitro methods. These replacements comprise excised animal tissues, which include bovine corneal opacity and permeability test and the isolated rabbit eye test. The EpiSkinA test developed by IMEDEX and L'Oreal has formally replaced the Draize test and has been approved by the European Union.

EpiSkin is human epidermis reconstructed in vitro and is derived from human keratinocytes. The cells are cultured at the air-1 interface on a collagen matrix. The cells are obtained from mammary samples during plastic surgery from healthy human donors with their informed consent. ECVAM approved the EpiDerm test as a full replacement of skin corrosion and skin irritancy test previously conducted in rabbits in 2007. Further in 2008, the SkinEthicRHE assay was endorsed by the Non-Commission members of ESAC. Both the models have been integrated in the test guideline 439 of the OECD. These in vitro skin models are being employed to screen and assess the chronic and acute skin irritancy of cosmetic as well as medical topical formulations as well as the parameters of phototoxicity. Few other commercially available skin models are EST 1000 SKIN MODEL (Cell Systems, Troisdorf Germany), EpiDerm Model (MatTekCorporation, MA, USA), The Straticell Model (Straticell, Les Isnes, Belgium), Labcyte Model (Gamagori, Japan), and Vitrolife Skin Model (Kyoto, Japan). The list of most routinely used alternative in vitro test systems and the OECD test guidelines have been given in Table 1.1 [11]. The detailed OECD guidelines are available at the site http://www.oecd.org/chemicalsafety/testing/oecdguidelinesforthetestingofchemicals.htm.

TABLE 1.1 Some Major Alternative In Vitro Testing Strategies and the Organization for Economic Cooperation and Development (OECD) Test Guidelines Governing Them

Genetic Toxicology		
Validated Alternative	OECD Test Guideline	Brief Description
Bacterial reverse mutation test (Ames test)	471	Uses amino acid requiring bacterial strains, to detect mutations induced by substances that restore the functional capability of the bacteria to synthesize an essential amino acid
In vitro mammalian chromosomal aberration test	473	Identifies agents that cause structural chromosome aberrations in cultured mammalian somatic cells
In vitro mammalian cell gene mutation test using the hprt or xprt locus	476	Detects gene mutations induced by chemical substances
In vitro mammalian cell micronucleus test	487	Detection of micronuclei in the cytoplasm of interphase cells as an endpoint of genotoxicity

TABLE 1.1 Some Major Alternative In Vitro Testing Strategies and the Organization for Economic Cooperation and Development (OECD) Test Guidelines Governing Them—cont'd

Genetic Toxicology		
Validated Alternative	OECD Test Guideline	Brief Description
Skin Irritation and Corrosion		
In vitro skin corrosion: human skin model test	431	RhE-human skin models: Episkin, Epiderm, SkinEthic EpiCS(EST-1000)
In vitro membrane barrier test method for skin corrosion	435	No EU test method available
Skin absorption: In vitro method	428	Provides information on the absorption of a substance applied to human and animal skin biopsies
Skin sensitization: local lymph node assay	429	Identifies contact allergens and skin sensitization hazards

THE AVAILABLE IN VITRO MODELS SYSTEMS

The debate between the accuracy of in vitro and in vivo is never ending; however, each has its pros and cons. In the upcoming sections, we shall discuss the types of in vitro systems and their potential applicability in the field of toxicology.

Explant/Organ Culture

Organ cultures are a connecting link between the in vivo and in vitro studies and are commonly referred to as the ex vivo studies. Organs are removed surgically and preserved to perform the in vitro assays later on. Organ culture is an age-old technique. Loeb in 1897 successfully sustained organs such as thyroid, ovary, kidney, and liver on small plasma clots for nearly 3 days. Tissue sectioning to obtain precision cut tissue slices (PCTS) is a step toward explants culture. Tissue sections from multiple organs can be prepared by sacrificing one animal thus allowing toxicology screening in multiple organs of same species in one go. Thus one can study the target organ toxicity as well as the cell–cell interactions via PCTS [12]. Tissue sections are important especially for imaging and immune-histochemical techniques, whereby allowing us to study the physiology, morphology, and cellular states of the treated animal-derived tissues.

Cell Culture

This is by far the most popular in vitro method, which comprises simple two dimensional (2D) monolayer cultures. Eukaryotic mammalian cells can easily be grown and expanded as either suspension or adherent cultures. The technique comprises the growth of isolated cells either of animal or human origin outside the body of live animals. The literal meaning "in-glass" perhaps comes from this very technique where cells either prokaryotic or eukaryotic are cultured in glass or plastic culture ware in humidified incubators provided with appropriate supply of carbon dioxide and oxygen. Detailed protocols and books are available for the various types of cell culture techniques. The cells are mainly of two types:

Primary Cell Culture

Here cells are isolated or harvested from fresh animal tissues such as the hematopoietic and mesenchymal stem cells derived from human umbilical cord blood, astrocyte, and glial culture from rodent models, neural stem cell culture from rat fetus, and so on. However, primary cultures are difficult to sustain because they have high metabolic and low proliferation rates. Early senescence and altered phenotypes are a few other hurdles. Primary cultures are also prone to contamination and require expertise of handling. However, if the tissue supplies are abundant these cultures are preferred over immortalized cell lines because they present a more chemical sensitive in vitro model system and the results are closer to the in vivo counterparts.

Immortalized Cell Lines

These are originally primary cells, which have been intentionally transformed or genetically modified to overpower the hurdles of primary cell cultures. The introduction of oncogenes enables rapid proliferation of cells, which are resistant to differentiation. They are passage stable and can be easily procured commercially. However, genetic manipulation and increased telomerase activity, senescence resistance, etc., make the cell lines devoid of original properties and thereby may not give an exact picture about the cellular responses. However, they serve as excellent in vitro screening tools and are widely used for simple toxicology studies.

Organotypic Cultures

These cell cultures are an attempt to mimic the in vivo heterogeneity of cells in vitro. The technique may be a simple 2D coculture of cells or a complex three-dimensional (3D) culture to mimic the in vivo niche like microenvironment. One of the best examples includes the human skin models. These are the classical cases of multicellular cultures which have yielded successful results in terms of cosmetic, drug testing, as well as transplantation cure purposes [13].

Stem Cells and Induced Pluripotent Stem Cells

Stem cells and induced pluripotent stem cells have given a new face to the field of in vitro toxicology. Stem cells are the naive cells of body with a capacity of self-renewal and unlimited proliferation and a unique ability to commit to a specific cell lineage. Biologists have successfully made use of all these characteristics of stem cells to deliberately direct them toward specific organ lineages and derive specific cells for various research purposes [14–16]. Directed conversion of stem cells into a specific lineage somewhat parallels the natural development process and has formed the basis of developmental toxicology studies. A sudden boom in stem cell technology has given us detailed protocols and methods, which are easy, reproducible, and universally accepted. The induced pluripotent stem cells (iPSC) technology is even ahead of simple stem cell technology by which simple somatic cells of the body can be reprogrammed into stem cells and can thus be used in the same way as the naïve stem cells. Patient-derived iPSCs are being used for the purpose of personalized toxicology screening and development of personalized medicine [17]. A detailed discussion about the stem cell approaches used for in vitro toxicology is beyond the scope of this chapter but we will briefly touch upon the applicability of the same in the upcoming sections of target organ toxicology.

Organoid Culture

The latest in vitro advancements have led us to the development of human organoid cultures which has brought scientists even closer to human-based research. Organoids are complex self-organizing multicellular constructs which are by far the closest possible mimics of an organ's in vivo like microenvironment known as the "niche". Human stem cells and iPSCs are being employed extensively to generate organoids, which find widespread applicability as potential in vitro tools for drug screening, toxicology testing, and even bioactivity of pharmaceuticals [18]. The organoid culture is the current and future face of in vitro toxicology and has sufficiently rid us of the moral burden of futile animal sacrifice. Except vascularization almost all the cellular, physiological, and molecular complexities of an intact organ have been recapitulated in these miniature organs.

BASAL TO ORGAN-SPECIFIC ENDPOINTS OF IN VITRO TOXICOLOGY

Basal Toxicity or Cytotoxicity

In vitro cytotoxicity studies have been employed for decades to assess the basal toxicity of various chemicals. These assays are simple, easy to perform, less time taking, and serve as the first basic step to screen the biological safe dose of a drugs and chemicals. These rely on single endpoints and are universal for all types of cells. A few examples of the popular cytotoxicity assays are as follows.

Tetrazolium Bromide Salt Assay

This assay is widely used and among the most popular colorimetric assays employed for the assessment of cell viability. The reduction of tetrazolium bromide salt by the mitochondrial dehydrogenase in living cells to a purple colored formazan precipitate, which on dissolution can be read spectrophotometrically in the visible region of light forms the principle of the assay.

Lactate Dehydrogenase Assay

The endpoint measured is membrane permeability. Membrane damage via toxic concentration of test agents results in the loss of intracellular lactate dehydrogenase, which can be measured spectrophotometrically in the culture medium. Apoptosis and necrosis are among the most widely studied objectives via this assay.

Neutral Red Uptake Assay

The lysosomal uptake of the supravital neutral red dye by living cells forms the principle of this assay. The lysosomal membrane damage is an indicator of toxic effect of chemicals and thus results in decreased uptake of the dye.

Trypan blue dye exclusion assay: This is one of the most conventional and oldest methods to assess cell viability. It is based on the simple principle of cell membrane integrity. Live cells with an intact membrane, exclude the trypan blue dye, whereas the dead cells with a damaged membrane easily uptake the dye and appear blue. This assay thereby gives the simplest yet reliable readings of total cell counts as well as percentage cell viability.

Acetoxymethyl Diacetylester of Calcein Cytotoxicity Assay

The assay is widely used to demonstrate and quantify cellular apoptosis. Acetoxymethyl diacetylester of calcein is a lipophilic dye taken up by viable cells where it is converted to calcein by intracellular esterases which gives a fluorescent green signal at 530 nm.

Adenosine Triphosphate Quantification and Adenosine Triphosphate/Adenosine Diphosphate Ratio Assay

Adenosine triphosphate (ATP) reflects the energy reservoir of living cells and its measurement serves as an important endpoint of cell viability. Toxic insult results in the drastic depletion of ATP, which can be quantified. The enzyme luciferase in presence of ATP, catalyses the conversion of luciferin to oxyluciferin which gives a luminescent signal. The ATP concentration is directly proportional to the signal intensity [19].

Genotoxicity

Genotoxicity is referred to as the ability of a physical or chemical agent to cause DNA damage, which results in a mutation. It is one of the most crucial endpoints of toxicology studies and helps in screening the preliminary potential hazards of any chemical or drug. Because the mechanism of action is at the cellular level, in vitro assays comprising isolated cell systems are more apt for the genotoxic studies. Mechanistic studies and xenobiotic metabolism can be better demonstrated in vitro and provides specific insights into a drug's target. A number of in vitro assays have been approved by the regulatory bodies and ICH has provided published guidelines for the same [20]. One of the most popular tools to assess the genotoxic effects in the form DNA damage and repair of prokaryotic and eukaryotic cells via chemicals and drugs is the "comet assay". The assay is a sensitive, rapid, and simple tool, which is widely applicable for the purpose of genotoxicology studies [21]. The popularity of this assay led to the development of the ComNet project, with a collaboration of nearly 100 research groups, which would share datasets for the effective assessment and monitoring of genotoxicity globally [22].

CELLS TO 3D ORGANOIDS: IN VITRO ORGAN SYSTEM TOXICOLOGY

Hepatoxicity In Vitro

Liver is the major organ responsible for drug/chemical metabolism and therefore is a prime target for all toxicology studies. Any organ-specific toxicology study cannot be carried out alone in vivo and requires in vitro techniques for understanding the cellular, metabolic, mechanistic, histopathological, and molecular aspects. Therefore, one of the simplest approaches is to treat animals with a sufficient dose of chemicals, sacrifice the animals, and surgically remove the organ to prepare either sections or primary culture of cells for further studies. Liver slices have served as the oldest models of hepatotoxicity as the histology remains intact, all the different cells can be observed and even the zonal distribution of cytochromes can be seen. Immortalized cell lines of hepatic origin such as Hep3B, Fa2N-4, HBG, HepG2, PLc/PRFs Huh7, and HepaRG are commercially available and are widely used. Although the phenotypic characters of the organ are lacking, the organ-specific genes are still present abundantly. The gene profile of the various phase I and phase II enzymes varies among different culture lots of different passages but these cell lines are by far the simplest and economical means to screen and assess the biological safety of potent hepatotoxins [23]. In vitro techniques have progressed substantially and a myriad of improvised protocols and models have come into picture. Stem cells and induced pluripotent stem cells have been successfully directed toward the hepatocytic lineages [24,25] and these stem cell derived hepatocytes serve as excellent and superior in vitro models of hepatotoxicity [26]. 3D organoid culture techniques are also being employed to mimic the in vivo like organ physiology in vitro and have met success. Earlier coculture of NIH-3T3 and HepG2 cells embedded in 3D hydrogel matrices were used to generate an array of hepatic organoids in a microfluidic-based approach. These organoids were superior from their 2D in vitro counterparts in having better cytochrome

P4503A4 activities and albumin secretion profiles and were successfully used to screen the toxicity of a potent hepatotoxin acetaminophen. These organoids were developed with the aim of "fail early, fail cheaply" principle, which allows for a quick, easy, time, and cost effective means of drug screening and hepatotoxicity testing [27]. However, now detailed protocol descriptions are available for the successful derivation of liver organoids from mouse as well as human stem cells. These organoids are genetically stable, retain their tissue origin, and are functionally active. Genetic manipulation via sophisticated molecular approaches further equips the in vitro toxicologists to efficiently employ these organoids for studying the various genetic and mechanistic aspects of drugs and chemicals [28].

Neurotoxicity In Vitro

Neurotoxicity and developmental neurotoxicity have been extensively studied via the in-vitro approaches, which have been revised from time to time. Perfused rodent brains, which are sectioned for morphological and histopathological studies, have given us ample insight into the various toxicological aspects of the central nervous system. Post-mortem human brain tissues have given us the current understanding of the various neurodegenerative disorders. However, human fetal brain procurement is not only ethically restraining but practically unfeasible too thereby limiting the understanding of developmental toxicology. Immortalized cell lines and primary culture of neurons, astrocytes, and glial cells have served as excellent in vitro models of neurotoxicity [29]. The rat pheochromocytoma-derived cell line PC12 and the human neuroblastoma cell line SH-SY5Y are some of the most popular age old in vitro models of neurotoxicology [30,31]. The advent of stem cell and iPSCs technology has given new dimensions to the field and a plethora of published studies are available. Stem cells and iPSCs can be successfully committed to the neuronal lineages and are serving as potential in vitro toxicology screens [14,32]. Our research group too has employed human umbilical cord blood derived stem cells for the establishment of a fast, rapid, and reproducible in vitro tool for assessing the toxic and mechanistic aspects of potent developmental neurotoxins such as monocrotophos and 3-methylcholanthrene [33–36]. Organoid research too has advanced considerably and generation of specific regions of the brain has met with substantial success. The group of Eiraku et al. has given us successful leads in the direction of embryonic stem cells' derived self-organizing neural structures of ectodermal origin since then have come up with several published protocols [37,38]. Human embryonic stem cells have been used for the generation of the dorsomedial telencephalic tissue to generate hippocampal neurons, thereby recapitulating hippocampus development successfully [39]. Methods for the generation of human midbrain like organoids have been standardized successfully, these organoids comprise the dopaminergic neurons as well as express neuromelanin producing neurons

[40]. Protocols and methods for the generation of retinal tissue and its successive transplantation have also been reported repeatedly [41–43]. Cerebral organoids too are being looked upon as potential in vitro tools to study and understand the complex process of brain development as well as the associated diseases [44]. However, generation of a complete brain organoid is still an unmet goal but is being worked upon extensively. Till then these region-specific organoids are serving as excellent tools for the purpose of neurotoxicity as well as developmental neurotoxicity.

Nephrotoxicity In Vitro

Kidney is a vital organ responsible for the fluid homeostasis of the body and is the basis for all the excretory functions. Along with the hepatic system, the renal system plays a crucial role in drug/chemical metabolism and the subsequent elimination of the metabolites from the body. Similarly, kidneys are prone to toxic insults by various chemicals and drugs, which can bring about fatal renal toxicity. The earliest in vitro models of nephrotoxicity comprised the PCTS prepared from rabbits as well as the primary culture of proximal convoluted and renal proximal straight tubules maintained in suspension. These served as the preliminary models for the assessment and screening of potent nephrotoxicants [45]. The renal system comprises typical cells committed for the various specific functions and therefore culture of the different cells is imperative for the understanding of the different aspects of renal toxicity. For example, culture of the mesangial cells and glomeruli in vitro has given us a clearer understanding of the glomerular function and the role of toxicant induced mesangial contraction in the functioning of the renal system. The mechanistic aspects of potent nephrotoxicants such as cadmium, cyclosporine, gentamicin, and cisplatin have been unraveled via these in vitro models of nephrotoxicity [46]. Development of novel cell lines such as the ones derived from renal proximal tubule epithelial cells is serving as better predictive models of nephrotoxicity as well as cancer studies. These cell lines are devoid of the chromosomal abnormalities and have been immortalized via the catalytic subunit of the human telomerase reverse transcriptase, which renders the cell line more sensitive than the other conventional ones [47]. Rat kidney cell lines such as the NRK-52E have been employed to understand and unravel the signaling cascades and mechanistics involved in the analgesic (acetaminophen) induced nephrotoxicity. The potential nephrotoxic effects via oxidative damage and subsequent protection via bioflavonoid Morin have been successfully demonstrated via this validated in vitro model [48]. Protocols for the development of functional kidney organoids derived from human pluripotent stem cells have been well standardized and published. Recapitulating the renogenic events via these self-aggregating and assembling structures will give us a deeper and more profound understanding of the various developmental processes, renal dysfunctions, signaling cascades, as well as nephrotoxicity [49,50].

Cardiotoxicity In Vitro

The preliminary ex vivo models of cardiotoxicity comprised perfused rat hearts, which were used to assess the toxic potential and mechanism of action of potent cardiotoxins (allylamine) as a simple means to demonstrate acute cardiotoxicity [51]. Cardiac toxicity studies have largely benefited from the progress of stem cell and iPSCs technology, which has given us the present day in vitro models of the same. Stem cells and iPSCs can be successfully committed toward the cardiac lineage and have been used to generate functional cardiomyocytes of human origin. These can be further used to screen and study the various molecular mechanisms, underlying pathways and even cancer induction by potential cardio-toxic drugs and chemicals [52–54]. Stem cell biology equipped with nanotechnology and material science has led to the development of engineered 3D organ tissues such as the heart tissue that finds widespread application in transplantation as well as drug toxicity screening. Further patient derived iPSCs are being used to construct such engineered heart tissues, which may serve the purpose of personalized medicine, therapeutic toxicity screening, and individual specific custom made drug strategies [55,56].

Nanotoxicology In Vitro

Nanotechnology has given new dimensions to the industrial sector with the manufacture of many consumer products as well as sensing devices. More than 20 countries globally are involved with the manufacture and marketing of nanotech-based commercial products, cosmetics being the largest category. With the advent of the technology has arisen the issue of nanotoxicology, which deals with the biosafety assessment and toxic responses of the nonmaterial and nanoparticles exerted both on living beings and the environment. Characterization of nanoparticles followed by detailed toxicity/biosafety profiling is as important aspect of the field. Nanoparticles (NPs) are extremely small in size and thereby easily cross-cellular barriers and also interact with DNA and proteins. Because the impact is at the cellular and molecular level nanotoxicity cannot be restricted to any one organ system and therefore various in vitro models have been employed for assessment of the same.

Metal nanoparticles such as titanium dioxide are largely present in cosmetic products such as sunscreens and lotions and are being manufactured in bulk globally. The human keratinocyte cell line (HaCaT) has been employed as an in vitro model for screening the cellular uptake and consequent assessment of cytotoxicity potential of the titanium dioxide nanoparticles (TiO_2 NPs) [57]. HepG2 (the human hepatocarcinoma cell line) has been used to assess the toxic potential of titanium dioxide NPs at extremely low concentrations such 1 µg/mL. This human liver-based in vitro model successfully demonstrated the DNA damaging role of TiO_2 NPs along with their ability to trigger the apoptotic events [58]. Similarly the human lung carcinoma cell line A549 too has been used as an in vitro screening model to assess the genotoxicity and DNA damaging potential of the TiO_2 NPs [59]. Primary culture of mice macrophages has been used an in vitro model to unravel the signaling and mechanistic aspects of the zinc oxide nanoparticles [60]. Recent advances in the field of in vitro nanosafety have led to the assessment of genotoxicity potential of NPs via 3D in vitro skin models such as the EpiDerm [61]. Comparative analysis between 2D and 3D models has clearly reflected the superior nature of the latter in predicting more in vivo like responses [61].

SUMMARY

The advent of novel technologies and research approaches has revolutionized the field of in vitro toxicology immensely. That which seemed almost an unachievable goal almost half a century back stands nearly accomplished today. Driven by feelings of empathy and compassion the scientific and medical fraternity across the globe has largely compensated for the indiscriminate animal sacrifice conducted over the years by moving toward suitable alternatives in the form of in vitro strategies. In vitro toxicology is almost a science in itself today and has progressed tremendously over the years. Spanning over more than five decades, in vitro strategies began with simple cell culture techniques and have gradually progressed to the level today where we have advanced stem cell-based 3D human organoids. We have partially won the battle for a more humane and ethical in vitro approach which can be directly extrapolated to human beings and we are gradually heading toward an era of more practical human-based in vitro toxicology. Such rapid, reproducible, cost-effective and high-throughput in vitro models of toxicology will not only yield valid results, will be instrumental in reducing clinical trial failure and will also rid us of the sociomoral burden of animal euthanasia.

SUGGESTED WEB PORTALS FOR FURTHER READING

1. http://www.dimdi.de/static/en/db/dbinfo/zt00.htm
2. http://www.gopubmed.org/web/go3r/WEB10O00f01000j100200010
3. http://searchguide.ccac.ca/index.php?id=3
4. www.frame.org.uk/page.php?pg_id=18
5. http://alttox.org/ttrc/emerging-technologies
6. www.wc8.ccac.ca
7. www.aimgroup.eu/2009/wc7
8. http://www.lasa.co.uk/Guidance%20notes%20RR%20(2004).pdf
9. http://ec.europa.eu/enterprise/epaa/index_en.htm
10. http://www.rspca.org.uk/sciencegroup/researchanimals/ethicalreview/retrospectivereview
11. http://www.skinethic.com/invitro.asp
12. ALTEX (Alternatives to Animal Experimentation, CH)—www.altex.ch
13. Altweb (Alternatives to Animal Testing on the Web, USA)—http://altweb.jhsph.edu

14. AVAR (Association of Veterinarians for Animal Rights, USA)—www.avar. org
15. AWIC (Animal Welfare Information Center, USA)—http://netvet.wustl. edu/awic.htm
16. CAAT (Center for Alternatives to Animal Testing, USA)—http://caat. jhsph.edu/
17. ECVAM (ECVAM Scientific Information Service, EU)—http://ecvam–sis. jrc.it/
18. FBL3R (Foundation Biographics Laboratory 3R, CH)—http://www. biograf.ch/
19. FFVFF (Foundation for Animal Free Research, CH)—http://www.ffvff.ch
20. FR3R (Foundation Research 3R, CH)—http://www.forschung3r.ch/
21. FRAME http://www.frame.org.uk/index.htm
22. HSUS (Humane Society of the United States, USA)—http://www.hsus.org
23. NCA (The Netherlands Center Alternatives to Animal Use)—http://www. nca-nl.org/
24. NORINA (Norwegian Inventory of Alternatives)—http://oslovet.veths.no/ norina/
25. SFRWAE http://www.algonet.se/
26. SET http://www. tierversuche-ersatz.de
27. UCCAA –http://www.vetmed.ucdavis.edu/Animal_Alternatives/main.htm
28. ZEBET   http://www.bgvv.de/cms/detail. php?template=internet_de_index_js
29. ZET (Center for the Replacement of Animal Experimentation, A)—http:// www.zet.or.at/
30. http://www.tga.gov.au/docs/html/euguideh.htm
31. http://www.emea.eu.int/pdfs/human/swp/259202en.pdf
32. www.animalethics.org.au
33. http://www.apvma.gov.au/guidelines/requirements3ag.shtm
34. www.endeuanimaltests.org
35. http://oberon.sourceoecd.org/vl=5003995/cl=13/nw=1/rpsv/cw/vhosts/ oecdjournals/1607310x/v1n4/contp1-1.htm
36. http://iccvam.niehs.nih.gov
37. http://www.who.int/ipcs/en/
38. https://www.epa.gov/

REFERENCES

[1] Hodgson E. Introduction to toxicology. Textb Mod Toxicol 2004:1.
[2] Tyson CA, Frazier JM. In vitro biological systems: methods in toxicology. Elsevier; 2016.
[3] Russell WMS, Burch RL, Hume CW. The principles of humane experimental technique. 1959.
[4] Kandárová H, Letašiová S. Alternative methods in toxicology: pre-validated and validated methods. Interdiscip Toxicol 2011;4(3):107–13.
[5] Pereira S, Tettamanii M. Ahimsa and alternatives-the concept of the 4th R. The CPCSEA in India. Altex 2005;22(1/05):3.

[6]   Worth AP, Balls M. The role of ECVAM in promoting the regulatory acceptance of alternative methods in the European Union. European Centre for the Validation of Alternative Methods. Altern Lab Anim ATLA 2000;29(5):525–35.

[7]   De Wever B, Fuchs HW, Gaca M, Krul C, Mikulowski S, Poth A, et al. Implementation challenges for designing integrated in vitro testing strategies (ITS) aiming at reducing and replacing animal experimentation. Toxicol In Vitro 2012;26(3):526–34.

[8]   Anadón A, Martínez MA, Castellano V, Martínez-Larrañaga MR. The role of in vitro methods as alternatives to animals in toxicity testing. Expert Opin Drug Metabol Toxicol 2014;10(1):67–79.

[9]   Fernando RN, Chaudhari U, Escher SE, Hengstler JG, Hescheler J, Jennings P, et al. "Watching the Detectives" report of the general assembly of the EU project DETECTIVE Brussels, November 24–25, 2015. Archiv Toxicol 2016;90(6):1529–39.

[10]   Rovida C, Martin SF, Vivier M, Weltzien HU, Roggen E. Advanced tests for skin and respiratory sensitization assessment. ALTEX 2012;30(2):231–52.

[11]   Buschmann J. The OECD guidelines for the testing of chemicals and pesticides. Teratog Test Methods Protoc 2013:37–56.

[12]   McKim J, James M. Building a tiered approach to in vitro predictive toxicity screening: a focus on assays with in vivo relevance. Comb Chem High Throughput Screen 2010;13(2):188–206.

[13]   Astashkina A, Mann B, Grainger DW. A critical evaluation of in vitro cell culture models for high-throughput drug screening and toxicity. Pharmacol Ther 2012;134(1):82–106.

[14]   Cao WS, Livesey JC, Halliwell RF. An evaluation of a human stem cell line to identify risk of developmental neurotoxicity with antiepileptic drugs. Toxicol In Vitro 2015;29(3):592–9.

[15]   Kang K-S, Trosko JE. Stem cells in toxicology: fundamental biology and practical considerations. Toxicol Sci 2011;120(Suppl. 1):S269–89.

[16]   Sánchez Alvarado A, Yamanaka S. Rethinking differentiation: stem cells, regeneration, and plasticity. Cell 2014;157(1):110–9.

[17]   Takahashi K, Yamanaka S. Induced pluripotent stem cells in medicine and biology. Development 2013;140(12):2457–61.

[18]   Ranga A, Gjorevski N, Lutolf MP. Drug discovery through stem cell-based organoid models. Adv Drug Deliv Rev 2014;69:19–28.

[19]   Mahto SK, Chandra P, Rhee SW. In vitro models, endpoints and assessment methods for the measurement of cytotoxicity. Toxicol Environ Health Sci 2010;2(2):87–93.

[20]   Suber R. In: Wallace HA, editor. Principles and methods of toxicology. New York: Raven Press Ltd.; 1994.

[21]   Dhawan A, Bajpayee M, Parmar D. Comet assay: a reliable tool for the assessment of DNA damage in different models. Cell Biol Toxicol 2009;25(1):5–32.

[22]   Collins A, Koppen G, Valdiglesias V, Dusinska M, Kruszewski M, Møller P, et al. The comet assay as a tool for human biomonitoring studies: the ComNet project. Mutat Res Rev Mutat Res 2014;759:27–39.

[23]   Soldatow VY, LeCluyse EL, Griffith LG, Rusyn I. In vitro models for liver toxicity testing. Toxicol Res 2013;2(1):23–39.

[24]   Huang P, Zhang L, Gao Y, He Z, Yao D, Wu Z, et al. Direct reprogramming of human fibroblasts to functional and expandable hepatocytes. Cell Stem Cell 2014;14(3):370–84.

[25]   Duan Y, Ma X, Zou W, Wang C, Bahbahan IS, Ahuja TP, et al. Differentiation and characterization of metabolically functioning hepatocytes from human embryonic stem cells. Stem Cells 2010;28(4):674–86.

[26]   Gómez-Lechón MJ, Tolosa L. Human hepatocytes derived from pluripotent stem cells: a promising cell model for drug hepatotoxicity screening. Archiv Toxicol 2016;90(9):2049–61.

[27] Au SH, Chamberlain MD, Mahesh S, Sefton MV, Wheeler AR. Hepatic organoids for microfluidic drug screening. Lab Chip 2014;14(17):3290–9.

[28] Broutier L, Andersson-Rolf A, Hindley CJ, Boj SF, Clevers H, Koo B-K, et al. Culture and establishment of self-renewing human and mouse adult liver and pancreas 3D organoids and their genetic manipulation. Nat Protoc 2016;11(9):1724–43.

[29] Schmidt BZ, Lehmann M, Gutbier S, Nembo E, Noel S, Smirnova L, et al. In vitro acute and developmental neurotoxicity screening: an overview of cellular platforms and high-throughput technical possibilities. Archiv Toxicol 2016:1–33.

[30] Greene LA, Tischler AS. Establishment of a noradrenergic clonal line of rat adrenal pheochromocytoma cells which respond to nerve growth factor. Proc Natl Acad Sci 1976;73(7):2424–8.

[31] Biedler JL, Helson L, Spengler BA. Morphology and growth, tumorigenicity, and cytogenetics of human neuroblastoma cells in continuous culture. Cancer Res 1973;33(11):2643–52.

[32] Bosnjak ZJ. Developmental neurotoxicity screening using human embryonic stem cells. Exp Neurol 2012;237(1):207–10.

[33] Kashyap MP, Kumar V, Singh AK, Tripathi VK, Jahan S, Pandey A, et al. Differentiating neurons derived from human umbilical cord blood stem cells work as a test system for developmental neurotoxicity. Mol Neurobiol 2015;51(2):791–807.

[34] Kumar V, Jahan S, Singh S, Khanna V, Pant A. Progress toward the development of in vitro model system for chemical-induced developmental neurotoxicity: potential applicability of stem cells. Archiv Toxicol 2015;89(2):265–7.

[35] Singh AK, Kashyap MP, Kumar V, Tripathi VK, Yadav DK, Khan F, et al. 3-Methylcholanthrene induces neurotoxicity in developing neurons derived from human CD_{34+} Thy_{1+} stem cells by activation of aryl hydrocarbon receptor. Neuromol Med 2013;15(3):570–92.

[36] Singh S, Srivastava A, Kumar V, Pandey A, Kumar D, Rajpurohit C, et al. Stem cells in neurotoxicology/developmental neurotoxicology: current scenario and future prospects. Mol Neurobiol 2015:1–12.

[37] Eiraku M, Sasai Y. Mouse embryonic stem cell culture for generation of three-dimensional retinal and cortical tissues. Nat Protoc 2012;7(1):69–79.

[38] Eiraku M, Sasai Y. Self-formation of layered neural structures in three-dimensional culture of ES cells. Curr Opin Neurobiol 2012;22(5):768–77.

[39] Sakaguchi H, Kadoshima T, Soen M, Narii N, Ishida Y, Ohgushi M, et al. Generation of functional hippocampal neurons from self-organizing human embryonic stem cell-derived dorsomedial telencephalic tissue. Nat Commun 2015:6.

[40] Jo J, Xiao Y, Sun AX, Cukuroglu E, Tran H-D, Göke J, et al. Midbrain-like organoids from human pluripotent stem cells contain functional dopaminergic and neuromelanin-producing neurons. Cell Stem Cell 2016;19(2):248–57.

[41] Nakano T, Ando S, Takata N, Kawada M, Muguruma K, Sekiguchi K, et al. Self-formation of optic cups and storable stratified neural retina from human ESCs. Cell Stem Cell 2012;10(6):771–85.

[42] Assawachananont J, Mandai M, Okamoto S, Yamada C, Eiraku M, Yonemura S, et al. Transplantation of embryonic and induced pluripotent stem cell-derived 3D retinal sheets into retinal degenerative mice. Stem Cell Rep 2014;2(5):662–74.

[43] Shirai H, Mandai M, Matsushita K, Kuwahara A, Yonemura S, Nakano T, et al. Transplantation of human embryonic stem cell-derived retinal tissue in two primate models of retinal degeneration. Proc Natl Acad Sci 2016;113(1):E81–90.

[44] Mason JO, Price DJ. Building brains in a dish: prospects for growing cerebral organoids from stem cells. Neuroscience 2016;334:105–18.

[45] Ruegg CE. Preparation of precision-cut renal slices and renal proximal tubular fragments for evaluating segment-specific nephrotoxicity. J Pharmacol Toxicol Methods 1994;31(3):125–33.

[46] Rodriguez-Barbero A, L'azou B, Cambar J, Lopez-Novoa J. Potential use of isolated glomeruli and cultured mesangial cells as in vitro models to assess nephrotoxicity. Cell Biol Toxicol 2000;16(3):145–53.

[47] Simon-Friedt BR, Wilson MJ, Blake DA, Yu H, Eriksson Y, Wickliffe JK. The RPTEC/TERT1 cell line as an improved tool for in vitro nephrotoxicity assessments. Biol Trace Element Res 2015;166(1):66–71.

[48] Mathur A, Rizvi F, Kakkar P. PHLPP2 down regulation influences nuclear Nrf2 stability via Akt-1/Gsk3β/Fyn kinase axis in acetaminophen induced oxidative renal toxicity: protection accorded by morin. Food Chem Toxicol 2016;89:19–31.

[49] Takasato M, Pei XE, Chiu HS, Little MH. Generation of kidney organoids from human pluripotent stem cells. Nat Protoc 2016;11(9):1681–92.

[50] Benedetti V, Brizi V, Xinaris C. Generation of functional kidney organoids in vivo starting from a single-cell suspension. Methods Mol Biol (Clifton, NJ) 2016.

[51] Anderson PG, Digerness S, Sklar J, Boor P. Use of the isolated perfused heart for evaluation of cardiac toxicity. Toxicol Pathol 1990;18(4 Pt. 1):497–510.

[52] Dick E, Rajamohan D, Ronksley J, Denning C. Evaluating the utility of cardiomyocytes from human pluripotent stem cells for drug screening. Biochem Soc Trans 2010;38(4):1037–45.

[53] Mandenius CF, Steel D, Noor F, Meyer T, Heinzle E, Asp J, et al. Cardiotoxicity testing using pluripotent stem cell-derived human cardiomyocytes and state-of-the-art bioanalytics: a review. J Appl Toxicol 2011;31(3):191–205.

[54] Mordwinkin NM, Burridge PW, Wu JC. A review of human pluripotent stem cell-derived cardiomyocytes for high-throughput drug discovery, cardiotoxicity screening, and publication standards. J Cardiovasc Translation Res 2013;6(1):22–30.

[55] Zhao Y, Feric NT, Thavandiran N, Nunes SS, Radisic M. The role of tissue engineering and biomaterials in cardiac regenerative medicine. Can J Cardiol 2014;30(11):1307–22.

[56] Tzatzalos E, Abilez OJ, Shukla P, Wu JC. Engineered heart tissues and induced pluripotent stem cells: macro-and microstructures for disease modeling, drug screening, and translational studies. Adv Drug Deliv Rev 2016;96:234–44.

[57] Shukla RK, Kumar A, Pandey AK, Singh SS, Dhawan A. Titanium dioxide nanoparticles induce oxidative stress-mediated apoptosis in human keratinocyte cells. J Biomed Nanotechnol 2011;7(1):100–1.

[58] Shukla RK, Kumar A, Gurbani D, Pandey AK, Singh S, Dhawan A. TiO_2 nanoparticles induce oxidative DNA damage and apoptosis in human liver cells. Nanotoxicology 2013;7(1):48–60.

[59] Kansara K, Patel P, Shah D, Shukla RK, Singh S, Kumar A, et al. TiO_2 nanoparticles induce DNA double strand breaks and cell cycle arrest in human alveolar cells. Environ Mol Mutagen 2015;56(2):204–17.

[60] Roy R, Singh SK, Chauhan L, Das M, Tripathi A, Dwivedi PD. Zinc oxide nanoparticles induce apoptosis by enhancement of autophagy via PI3K/Akt/mTOR inhibition. Toxicol Lett 2014;227(1):29–40.

[61] Wills JW, Hondow N, Thomas AD, Chapman KE, Fish D, Maffeis TG, et al. Genetic toxicity assessment of engineered nanoparticles using a 3D in vitro skin model (EpiDerm™). Part Fibre Toxicol 2016;13(1):50.

Chapter 2

Principles for In Vitro Toxicology

Sonal Srivastava, Sakshi Mishra, Jayant Dewangan, Aman Divakar,
Prabhash K. Pandey, Srikanta K. Rath
CSIR-Central Drug Research Institute, Lucknow, India

INTRODUCTION

The 3 Rs principles describe reduction, refinement, and replacement alternatives to animal testing in order to bridge the gap between the animal-based in vivo toxicity screening techniques and the new cell and tissue-based methods. Although the prime focus of this series is on replacement alternatives, an introduction to the 3 Rs is required for better understanding of the principles of in vitro toxicology.

- The first R, reduction, refers to the use of fewer animals by statistically designing the study in order to obtain data that can be compared with the results obtained from a study using large number of animals. Furthermore, the data gaps resulting from the reduction of animals should be complemented by including in vitro end points.
- The second R, refinement, emphasizes on reducing the suffering and pain of the animal. Refinement also comprises the use of species located lower on the phylogenic scale. Both reduction and replacement of animals can be achieved by adopting refinement techniques.
- The third R, replacement, focuses on the use of various nonanimal techniques based on established cell lines, proliferative cell cultures, and tissues which can reduce the number of animals needed for completion of a study [1].

Chemical, pharmaceutical, cosmetic industries, and regulatory bodies routinely use in vitro methods for toxicity testing, safety assessment, and risk evaluation. The cellular and molecular events occurring in vitro can be correlated with the in vivo physiological reactions and can help in understanding why a specific in vivo response has been obtained in one species and may not be seen in another. Therefore, selection of an appropriate in vitro assay allows better prediction of an outcome in vivo and also offers the ability to investigate a number of specific events and end points. The objective of this chapter is to highlight the principles and procedures that have particular relevance in the field of in vitro toxicology.

In Vitro Toxicology. http://dx.doi.org/10.1016/B978-0-12-804667-8.00002-X

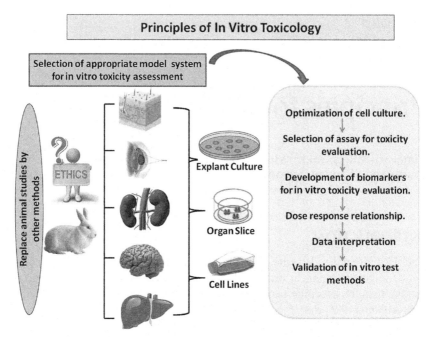

FIGURE 2.1 Principles of in vitro toxicology: an overview of in vitro systems used for toxicity assessment and replacement of animal models.

The emphasis will be more on the in vitro models and their application for toxicity evaluation. An overview of the protocols specific to individual tissue types has also been presented in this chapter. Fig. 2.1 sheds light on the spectrum of in vitro models that replace the animal models for toxicity assessment and the basic principles that help to obtain reliable and reproducible data from these test systems.

GUIDELINES GOVERNING THE IN VITRO METHODS FOR TOXICITY TESTING

With the rising ethical concerns on animal experimentation, most regulatory authorities have shown their commitment to adopt the 3Rs principle which recommends the reduction in the use of animals in toxicology testing and replacing them with appropriate in vitro tests. Recent major developments have led to the incorporation of in vitro systems, not only in basic research, but also to meet regulatory requirements for toxicity testing of chemicals and quality control of various products. In order to maximize the reliability, reproducibility, credibility and acceptance of the result obtained from in vitro procedures, it is essential to focus on the regulatory aspects of development of in vitro methods.

Good Cell Culture Practices (GCCP) are regulatory guidelines which attempt to define minimum common standards in cell and tissue culture that will reduce uncertainty in the development and application of in vitro procedures. This GCCP guidance lists a set of the following six principles intended to promote consensus among all concerned with the use of cell and tissue systems [2].

1. establish and maintain best laboratory practices
2. promote effective quality control systems
3. facilitate education and training
4. support journal editors and editorial boards
5. assist research funding bodies
6. Facilitate the interpretation and application of conclusions based on in vitro data.

The Organisation for Economic Co-operation and Development (OECD) has framed Test Guidelines (TG) that will expand options for the use of in vitro methods in the areas of skin and eye irritation, and genotoxicity. Following is a list of these guidelines, with brief descriptions of their applicability.

TG 491: *Reconstructed Human Cornea-like Epithelium (RHCE) Test Method for Identifying Chemicals not Requiring Classification and Labelling for Eye Irritation or Serious Eye Damage.* This is the first guideline for ocular testing to use reconstructed human tissue rather than ex vivo animal tissue. It is based on the EpiOcular™ Eye Irritation Test (EIT) to assess chemicals that do not require classification according to the UN Globally Harmonized System of Classification and Labelling of Chemicals (GHS). The test is to be used in combination with other recommended alternatives, in order to replace the Draize rabbit eye test [3].

TG 442D: *In vitro Skin Sensitisation: ARE-Nrf2 Luciferase Test.* This test method makes use of an immortalised adherent cell line derived from HaCaT human keratinocytes stably transfected with a selectable plasmid for measuring gene expression in ARE-dependent pathways, one of the indicators of the second key event in the skin sensitization Adverse Outcome Pathway [4].

TG 430: *In vitro Skin Corrosion: Transcutaneous Electrical Resistance Test Method (TER).* This test method utilizes skin discs taken from humanely killed rats to identify corrosives by their ability to produce a loss of normal stratum corneum integrity and barrier function [5].

TG 431: *In vitro Skin Corrosion: Reconstructed Human Epidermis (RHE) Test Method.* The test material (solid or liquid) is applied uniformly and topically to a three-dimensional human skin model, comprising at least a reconstructed epidermis with a functional stratum corneum. The principle of the human skin model assay is based on the hypothesis that corrosive chemicals are able to penetrate the stratum corneum by diffusion or erosion, and are cytotoxic to the underlying cell layers [6].

TG 435: *In vitro Membrane Barrier Test Method for Skin Corrosion.* This test method detects membrane barrier damage caused by corrosive test substances after the application of the test substance to the surface of the artificial membrane barrier, presumably by the same mechanisms of corrosion that operate on living skin [7].

TG 439: *In vitro Skin Irritation – Reconstructed Human Epidermis (RhE) Test Method.* The RhE test system uses human derived non-transformed keratinocytes that closely mimic the biochemical and physiological properties of the upper parts of the human skin, i.e. the epidermis. The test chemical is applied topically and the cell viability in RhE models is measured by enzymatic conversion of the vital dye MTT [8].

TG 428: *Skin Absorption: in vitro Method:* The test substance is applied to the surface of an excised skin sample separating the two chambers of a diffusion cell. The chemical remains on the skin for a specified time under specified conditions. At the end of the experiment the distribution of the test chemical and its metabolites are quantified [9].

TG 476: *In vitro Mammalian Cell Gene Mutation Tests using the Hprt and Xprt genes*: The cell lines used in these tests measure forward mutations in reporter genes, specifically the endogeneous hypoxanthine-guanine phosphoribosyl transferase gene and the xanthine-guanine phosphoribosyl transferase transgene (gpt) (referred to as the XPRT test). Cells in suspension or monolayer cultures are exposed to the test substance, both with and without an exogenous source of metabolic activation and then sub-cultured to determine cytotoxicity and to allow phenotypic expression prior to mutant selection. Following phenotypic expression, mutant frequency is determined by seeding known numbers of cells in medium containing the selective agent to detect mutant colonies, and in medium without selective agent to determine the cloning efficiency (viability) [10].

TG 492: *In vitro Mammalian Cell Gene Mutation Tests Using the Thymidine Kinase Gene.* Cells in suspension are exposed to the test substance, both with and without an exogenous source of metabolic activation. Cells deficient in thymidine kinase enzyme activity due to the mutation TK+/- to TK-/- are resistant to the cytostatic effects of the pyrimidine analog trifluorothymidine (TFT). The TK proficient cells are sensitive to TFT, which causes the inhibition of cellular metabolism and halts further cell division. Thus, mutant cells are able to proliferate in the presence of TFT and form visible colonies, whereas cells containing the TK enzyme are not [11].

TG 487: *In vitro Mammalian Cell Micronucleus Test.* Cell cultures of human or mammalian origin are exposed to the test substance both with and without an exogenous source of metabolic activation and analyzed for the presence of micronuclei in the cytoplasm of interphase cells [12].

TG 473: *In vitro Mammalian Chromosomal Aberration Test.* Cell cultures of human or other mammalian origin are exposed to the test chemical both with and without an exogenous source of metabolic activation. Cells are treated with Colchicine to induce metaphase arrest and analyzed microscopically for the presence of chromatid-type and chromosome-type aberrations [13].

OPTIMIZATION OF CELL CULTURE

Today, cell culture is a major tool for preclinical studies in life sciences. Optimization of cell culture needs a lot of time and meticulous efforts; therefore, at times many researchers spend much time to choose the correct cell type, growth medium, and culture configuration to achieve their goals. The term in vitro refers to the handling of biological specimen from outside the living system under controlled environment which support their growth, disparity, and solidity. In vitro cytotoxicological methods include the study of interaction of drugs and other chemicals with cells (human or animal origin), which cause the toxic effects. The continuous cell lines mostly used for research purpose have some characteristic features like chromosome heteroploidy, loss of contact inhibition, and ability to induce tumors in immune-compromised hosts and can also form colonies in soft agar. Some specific markers, which express in different cell lines are used to characterize and examine the cell culture progression for assessment of in vitro cytotoxicity [14].

There are several methods used for the identification, classification and establishment of the genetic purity of the cell lines' [15]. **Karyotyping** of derived cell lines are performed for chromosomal complementarity, i.e., chromosome of the cell lines and parent cell must be identical. Aging of cells in culture is measured in terms of the population doubling level (pdl), means number of times a cell undergoes mitosis since its isolation. The signs of aging like loss of cell shape and high content of cytoplasmic lipid are shown by finite cell lines. In case of finite cell lines, cell cycles are arrested in the Go phase after forming monolayer whereas continuous cell line do not form the contact inhibition and grow in multilayers. For toxicological observation the integrity of cell lines are determined by the measurement of growth and its viability by vital dye exclusion method. The certified cell banks, such as the European Collection of Animal Cell Cultures (ECACC, UK) and the American type culture collection (ATCC, USA) are the major sources of cell lines around the world. The certified cell lines from these sources have complete details about donor, karyotype analysis, expected population doubling time, cell structure features, and expression pattern of cell-specific markers.

There is a common risk of cross-contamination during cell culture handling especially with continuous cell lines and sterile conditions should be maintained to avoid this. The cross-contamination in cell lines is also detected by short tandem repeat (STR) profiling which is an efficient and reliable tool. STR profiling is also used for characterization of cell line or confirmation of cell line identity. Microbial contamination includes mainly bacteria, fungi, yeast, and mycoplasma. Among all, the mycoplasma contamination is most serious as it is not visible by routine microscopy. New cultures must be tested for microbial contamination on arrival and existing stocks should be monitored at regular intervals [16].

So far, methods for estimating the passage number of continuous cell lines do not exist. The continuous cell lines maintained in laboratories may differ by hundreds of passage. The cell lines do not retain similar characteristics with increased passage number. The long-term culturing or subculturing of cell line causes the divergent effects on cell morphology, development, and gene expression [17]. The use of animal serum for cell culture can also cause phenotypical differences in the cell cultures as it has an undefined composition that varies from batch-to-batch. This is due to the seasonal and continental variations to which the source of the serum is subjected to. It is advisable to conduct a set of experiments using serum from the same batch and commercial source. The optimization and implementation of the use of a chemically defined media is essential for avoiding variations in the output of in vitro test methods [18].

The general health of growing culture is evaluated based on three important cell characteristics, i.e., morphology, plating efficiency and growth rate. These features are also important in evaluating experimental results. It is easy to determine the **morphology** or cell shape but is often least useful. The morphology of cells changes frequently during culture and it is very difficult to relate the observations to the cause of the problem. Quantification or measurement of the precise morphology of cells is very difficult. The **growth rate** determination includes the estimation of cell count, which is highly sensitive to changing culture environment [19]. This helps in designing of experiments to determine best suited set of conditions, i.e., to achieve best growth rate in specified condition. These same techniques are useful for measuring cell survival or death in in vitro cytotoxicity assay. The small number of cells (20–200) is placed in culture vessel and the number of colonies they form is measured as **plating efficiency**. The survival is measured in form of the percentage of cell forming colonies, whereas size of the colony is the measure of growth rate.

MODELS OF IN VITRO TOXICITY

In the last decade many studies have focused on in vitro studies, as it provides an effective tool for toxicity evaluation and safety assessment. Increase in the complexity of in vitro models reflects its native in vivo environment, thereby easing the deduction of valid predictions from these test system [20]. In vitro models are designed in a way that they can be interrogated for numerous endpoints upon exposure to a toxicant and the result obtained is reproducible and easy to correlate to the toxic effects in the target organ.

Explant Culture

It is one of the oldest tissue culture techniques. Benefit of using explant culture for toxicological studies is that it holds histological architecture. It is widely used for study of carcinogens, environmental radiations or role of hormones. It has advantage in tissue engineering to mimic the organ-like structure. One more advantage of explant culture over total organ culture system is that it gives better visual results for experiments like immunohistochemistry as multiple precision-cut tissue slices can be obtain [21]. Explant culture of many types of organs such as liver, kidney, bladder, bronchus, esophagus, and skin has been successfully done. The drawback of this model is that the availability of nutrient and oxygen is limited to internal cells of explant and damage of few cells during explant slice preparation. These limitations can be surmounted by advanced tissue slice cutter which makes explant very thin and uniform.

Organotypic Culture

It is a kind of in vitro technique in which different cell lines are used to recapitulate cell heterogeneity in vitro. Mainly primary or continuous cell lines are used in this technique. This culture method incorporates supporting matrices to mimic organ culture or 3D scaffolds to create in vivo like environment which support tissue architecture and morphology. Organotypic culture is complex and multicellular in nature and is used to develop certain tissue models. A component of any organ is prepared by using combination of cells in a specific ratio which mimics the tissue in vivo conditions. Organotypic culture is a helpful tool in developing specific disease models for studying their respective targets. It also helps in assessment of both normal and diseased physiology of any diseased models. Among different models the skin model is a well-characterized organotypic culture system [22]. For example skin models are used in different pharmacological, toxicology and pharmacokinetics studies. Similarly for the function and dysfunction studies of central nervous system the hippocampal slice culture is widely used [23]. In addition, the oraganotypic culture is also used to screen compounds in absence of target toxicity information.

Cell Cultures

Use of cell cultures provides a better alternative model for the toxicity screening [24]. Cells of any species are easy to maintain and propagate without ethical concern. Less experimental variations in this system lead to high statistical significance, easy understanding of molecular mechanism, easy experimental setup, low experimental cost, and saves time. Drawback of using continuous cell lines in toxicity assessment is that it is a very simple system, whereas animal models are much more complex. Because of this limitation it can be used only for specific studies. Systemic toxicity assessment is not possible in cell lines; also it is difficult to determine the dose and to correlate the in vitro data with in vivo work. Though, continuous cells are *en vogue* because of its immortality feature, their use is restricted as they may differ in chromosome number and biochemical properties from the parent tissue. Primary cultures are comparatively more suitable for toxicity studies as it retains more characteristics of in vivo samples such as similar chromosomal numbers and the specialized biochemical properties, e.g., liver cells are able to secrete albumin [25]. Limitation in using primary cells is that it cannot run for a longer passage.

Three Dimensional Culture Systems

Due to the attributes like improved cell–cell interactions, cell–extra cellular matrix interactions; cell populations and structures that resemble in vivo architecture; 3D cell culture models are better models than the traditional 2D monolayer culture. 3D cell culture models can encompass healthy cells, diseased tissue biopsies or complete organ as their living components. 3D culture systems and cell culture scaffolds are ordinarily made by suspending cells in well matched hydrogels. These hydrogels can be mildly cross-linked in vitro to suspend cells within these materials in the presence of excess water, or by seeding cells on solid support matrices. Solid supports are typically spongy high-porosity polymers or ceramic solid foams, or growing cells on the surface of a nonplanar support material.

3D cell culture models have significance in the identification of molecular targets, regeneration studies, tissue engineering, disease modeling, in vivo toxicological studies, stem cell research, cancer cell biology, and in the validation of drug targets, etc. 3D cell culture models also provide more physiologically relevant information as well as more predictive data for in vivo tests [26].

DEVELOPMENT OF BIOMARKERS FOR IN VITRO TOXICITY EVALUATION

There is a need for the development of in vitro toxicity biomarkers for validation of clinical endpoints. An in vitro biomarker of toxicity provides a quantitative characteristic that serves as an indicator of an adverse effect and is mechanistically relevant to biochemical or molecular events in vivo. Establishment of appropriate in vitro systems as alternative models for risk evaluation can assist in prediction of toxic response in vivo. Oxidative stress, cellular stress responses, changes in enzyme activities, glutathione homeostasis, and cytokine responses are few early cellular events that occur in response to the treatment [27]. Genomics, transcriptomics, and proteomics approaches are employed to measure a multitude of endpoints as "biomarkers of effect" and "biomarkers of exposure, but not every endpoint predicts hazardous effects in vivo and qualifies as a biomarker of toxicity [28]. Relevance of the chosen in vitro approaches greatly determines their ability to be extrapolated to an in vivo context, therefore choice of what to measure becomes very important.

ASSAYS FOR SCREENING OF TOXICANTS

For the development of any drug toxicity assay, testing of new compounds is an essential step. The toxicity of any new compound can be assessed broadly in two ways: (1) by using cells/cell lines in in vitro studies; (2) by using experimental animals as in vivo exposure. Here, we are mainly focusing on the various in vitro culture methods used for toxicity testing of a given compound. Various assays used for in vitro toxicity evaluation have been discussed below and a brief illustration regarding the screening test and their end points has been summarized in Table 2.1.

Ames Test

DNA being chemically the same in all organisms, any living organism can be used to test for mutagens. Ames test is used for detection of mutagenic property of any drug or chemical. In this reverse mutation assay different strains of *Salmonella typhimurium*. It is the most accepted method for testing genotoxicity in vitro [29]. Different tester strains are having different types of mutations on histidine biosynthesis gene so that, it is unable to form colony in histidine deprived medium. Some strains like TA1535 have base pair substitution, while other strains like TA1537 and TA1538 have frame shift mutation. If these auxotrophic mutant bacteria are able to form colony under histidine negative medium in presence of test substances that shows mutagenic property of that compound. That mutation leads to conversion of auxotrophic bacteria into prototroph [30]. Ames Mutagenic Assay is routinely used in industry and regulatory toxicity studies of xenobiotics and drugs.

The Hen's Egg Test–Chorio Allantoic Membrane Test

The chick embryo chorio allantoic membrane (CAM) is a nonmammalian alternative model for toxicological testing of drugs. Chick CAM model system is an intermediate model between cellular and animal model as it gives tissue like response. The HET-CAM (Hen's Egg Test–Chorio Allantoic Membrane) test is an organotypical test in which test material is applied to the CAM of chicken eggs on day 9 of embryonation. The acute irritating effects induced by the chemical on the small vessels and protein membranes of this soft tissue membrane are comparable to the

TABLE 2.1 Various Methods, Models, and Their Endpoint for In Vitro Toxicity Assessment

Toxicity	In Vitro Test Method	Model	End Point	References
Nephrotoxicity	Epithelial barrier function measurement	LLC-PK1 monolayer cultures	Cell injury	[54]
	Alamar blue	Renal cortical slices, isolated nephron segments, primary renal cell cultures, renal cell lines—LLC-PK1, MDCK, LLC-RK1, and OK cells	Vital dye uptake	
Hepatotoxicity	Neutral red cytotoxicity	Liver slices, immortalized cell lines—HepG2, Hep3B, Huh7, primary hepatocyte suspension/culture, 3D cultures	Vital dye uptake or release indicating lysosomal activity or cell viability	[55]
Neurotoxicity	LDH assay	Organ cultures—spinal cord, ganglia, or sensory organs; explants culture—hippocampal slice culture, suspension, and reaggregate cultures, dissociated primary culture, continuous cell lines—neuroblastoma C1300 and NIE115 cells, C6 glioma cells, P12 pheochromocytoma cells	Release	[56]
	Trypan blue exclusion		Dye exclusion	
	MTT assay		Mitochondrial oxidation	
Cardiotoxicity	Measurement of organelle membrane integrity	Primary cultures of rat myocardial cells	Membrane fragility	[57]
	LDH leakage		Increased leakage	
	Calcium uptake		Decrease in uptake	
Pulmonary toxicity	MTT	Human fetal lung fibroblasts, such as HFL1, MRC-5, and WI38	Mitochondrial oxidation	[58]
	LDH		Release	
Immunotoxicity	MTT	Finite cell lines—human placental mast cells, thymocytes, and polymorphonuclear lymphocytes. Continuous cultures—RPMI 7666, RPMI 6666, human primary cells	Mitochondrial oxidation	[59]
	Human whole-blood cytokine release assay		IL-4 release suppressed	

Category	Test	Model	Endpoint	Ref.
Cutaneous irritant	ICE assay, Skin irritation test, Rat skin TER test, In vitro, percutaneous, absorption	3D reconstructed human tissue models—EpiDermTM and EpiSkin and the RhE, excised animal, or human skin	Tissue viability	[60]
Ocular irritant	HET-CAM test	Chorion allantoic membrane (CAM) of chicken eggs	Hemorrhage, lysis, and coagulation	[61]
	BCOP test	Bovine corneas	Opacity and permeability	
	RBC hemolysis test	Human erythrocytes	Membrane lysis and protein denaturation	
Genotoxicity 1. Gene mutation 2. Chromosomal abberations	Ames test, Metaphase analysis, FISH	*Salmonella typhimurium, Human lymphocytes, Chinese hamster ovary cell line*	DNA damage, Chromosome breakage or polyploidy	[30]
Embryotoxicity	Embryonic stem cell test (EST), FACS-ESTmethod	mouse embry-onic stem cells	Inhibition of differentiation into beating cardiomyocytes, stem cells cytotoxicity, 3T3 fibroblasts cytotoxicity	[62]
Teratogenicity	Micromass teratogen test	chick, mouse or rat embryo midbrain or limb cells	Loss of cell, adhesion, decreased division, and differentiation	[63]
Drug toxicity	Live/dead assays—Calcein AM/HoecHTS 33342, DAPI, and propidium iodine (PI), Caspase 3/7 assays	Primary cell culture, 2D and 3D cultures, organotypic culture, immortal cell lines like HEK 293T	Loss of, membrane integrity, apoptosis induction cytolysis	[64]

BCOP, Bovine Cornea Opacity/Permeability test; DAPI, 4′,6-diamidino-2-phenylindole; FISH, fluorescent in situ hybridization; HET, Hen's Egg Test; ICE, In Vivo Complex of Enzyme; LDH, lactate dehydrogenase; LLC, Lilly Laboratories Cell; PK, porcine kidney; RBC, red blood cell; RPMI, Roswell Park Memorial Institute medium; TER, trans-cutaneous electrical resistance.

effects on the eye. The end points of hemorrhage, lysis, and coagulation are assessed in an observation period of 5 min after the application of the chemical at the CAM [31].

Corrosivity Assays

Skin corrosion refers to the production of irreversible damage to the skin, manifested as visible necrosis through the epidermis and into the dermis, following the application of a test material. In vitro dermal corrosivity tests were developed as an alternative for the standard in vivo rabbit skin procedure.

1. **Membrane barrier test method:** This test method comprises of a synthetic macromolecular biobarrier to which the test substance is applied. Penetration of the membrane barrier (or breakthrough), caused by corrosive test substances, is detected by a chemical detection system having a pH indicator dye or combination of dyes, e.g., cresol red and methyl orange, that will show a color change in response to the damage [32].

2. **Transcutaneous electrical resistance (TER) test method:** This method uses the skin from humanely killed rat and the test chemical is applied to the skin discs. Corrosive substances are able to produce a loss of normal *stratum corneum* integrity and barrier function, which is measured as a reduction in the TER below a threshold level (5 kΩ for rat). A dye-binding step enables to determine whether the increase in ionic permeability is due to physical destruction of the *stratum corneum*.

3. **Reconstructed human epidermis test method:** This method uses a three-dimensional (3D) cell culture-based model comprising of nontransformed, human-derived epidermal keratinocytes, cultured to form a multilayered highly differentiated model of human epidermis. A corrosive chemical will be cytotoxic to the underlying layers upon penetration. The endpoint of this assay is an assessment of cell viability, determined by measuring the rate of conversion of MTT to its formazan salt by mitochondrial reductase enzymes [6].

3T3 Neutral Red Uptake Test

Reaction of the skin with any chemical and ultraviolet light or radiation exposure is termed as photosensitization. It can be a result of topical application of any chemical substances or exposure to radiation. 3T3 NRU phototoxicity test is one of the approach for in vitro phototoxicity testing. 3T3 mouse fibroblast cells are exposed to chemicals in presence or absence of UVA light. Cytotoxicity is estimated by neutral red dye uptake. Photoirritation can be measured by increase in cytotoxicity. This test is preferable over other tests because its data has high correlation with in vivo data [33].

Tetrazolium Salt, 3-(4,5-Dimethylthiazol-2-yl)-2,5-Diphenyltetrazolium Bromide Assay

The tetrazolium salt, MTT, is actively absorbed into cells and reduced in the mitochondria to yield a formazan product. The product cannot pass through the cell membrane and is liberated upon addition of a solvent like dimethyl sullfoxide, which is readily quantified colorimetrically [34]. The ability of the mitochondria to reduce MTT indicates mitochondrial integrity and serves as a measure of cell viability [35].

Alamar Blue

Alamar blue is an another tetrazolium salt that is converted to a fluorescent product but unlike MTT which deposits as an insoluble formazan, the reduced product of Alamar blue is freely soluble. It is a vital stain that has no adverse effects on cells. This property eases the assessment of mitochondrial and cyto-plasmic reduction over a period of time [36].

The Bovine Cornea Opacity/Permeability Test

The bovine corneas from eyes are used to assess the eye-irritating properties of test materials. The corneas are mounted in special holders having anterior and a posterior chambers filled with simple culture medium. Corneas are exposed for 1 h to the test material and a change in light transmission passing through the cornea is determined with an opacitometer. The permeability of the cornea is determined by exposure to fluorescein for 90 min [37].

DOSE–RESPONSE RELATIONSHIP

"In all things there is a poison, and there is nothing without a poison. It depends only upon the dose whether a poison is poison or not."—Paracelsus (1493–1541). Various features of exposure and the total spectrum of effects altogether in a correlative relationship are referred as dose–response (DR) relationship. A DR relationship defines the effect of toxicant on an organism caused by different levels of dose exposure [38]. In the last decade, there has been keen interest for studying the interactions between organisms and vari-ous environmental toxicants. There are several toxicants which even at lower concentrations are able to manifest toxicological effects. Therefore, the study of interactions among toxicants in in vitro models, particularly for molecular mechanisms of toxicity is an emerging field in the experimental toxicology. The interaction of toxicants in any organism might display various effects [39]. The cell can be simplified as a system having specific affinity for dif-ferent toxicants. Therefore, the interaction between the two may induces a quantifiable direct/indirect effect, which can be positively/negatively modi-fied by the presence of other toxicant. These effects are usually categorized into four groups. **Additive effect**: This is usually the combine effect of two toxicants where there effect is equal to the sum of effects of each agent given alone. When two or more than toxicants act together without any interac-tion amongst them and their overall effect is not different from concentration effect/response relationship of single compounds, is known as additive effect [40]. **Synergistic effect**: This is when the combine effects of two toxicants are greater than the sum of the effects of each toxicant given alone [41]. **Potentiation effect**: When one or more toxicant does not have a toxic effect on a certain system but when added with another may result in a pharmacologic response greater than the sum of individual responses to each toxicant [42]. **Antagonistic effect**: When the combine effects of two chemicals interfere with each other's action, which means that overall effect of two or more toxicants is less than the sum of their individual effects [43].

Evaluation of DR Relationship

The importance of dose effect/DR relationships in vitro experiments is to assess and to extrapolate the toxicological functions of any toxicants. In any in vitro system, the effects are usually measured are a change in cell viability, an increase of cell death, or a decrease in cell number (both via necrosis and apoptosis). The mathematical function that describes the same is represented by the DR curve Fig. 2.2, which mainly depends on the dose of toxicant. The DR curve is a helpful tool to understand the levels at which the toxicants begin to exert adverse effects on any cell system and the degree of harm expected at various levels [44,45]. DR curves can show a number of points including:

1. The "no effect level" where no effect is detectable with the given toxicant on any cell system.
2. The threshold dose of the substance—the level at which the effect starts to occur.
3. IC (inhibitory concentration) is the level where the effect occurs in a set percentage cells or all of the cells. IC_{50} (inhibitory concentration, 50%) is the concentration of any toxicant that kills 50% of the test population.

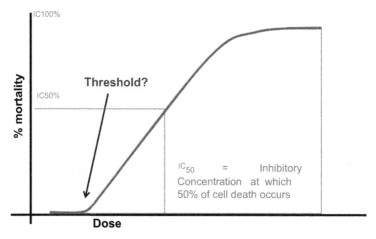

FIGURE 2.2 Dose–response curve.

In any experiment, when a specific is conducted for toxicity assessment of any toxicant at a given concentration, it becomes essential to get DR curve describing the observed response. The mathematical DR curve is drawn extrapolating the data of toxic effect for any toxicant to study the interactions [46]. There are essentially two possibilities for performing the studies: (1) once conducting any experiment with "N" number of replicates for each as well as different time point; (2) performing the experiments more than one time, with a relatively lower number of replicates [47]. Therefore, maintaining the optimum conditions for any experimental setup and making sure about the highest experimental repeatability and accuracy becomes an essential requirement. Meticulous, minimum concentrations of the toxicant, and the duration of exposure should also be checked accurately.

DATA INTERPRETATION

Data interpretation is an important factor in biological research as its leads better understanding and new finding for any result. Arrangement of all data in an organized way is the key step for data interpretation. Every quantitative data needs to be statistically analyzed. The statistical methods aid in data interpretation. There are many computational software for data interpretation like as Statistical Package for the Social Sciences (SPSS) or analysis of variance (ANOVA) for analyzing data [48]. Statistically analyzed qualitative data help in measuring the fold change in any cell system and also help in predicting the consistency of data. Statistical measures such as "significant" test results and P values always need interpretation. Most of the in vitro test results are culminated in terms of the P value. The P-value is commonly defined as the probability of getting a result (a test statistic) that is at least as extreme or equal to than what was actually observed, assuming that the null hypothesis was true. P-value is widely used in hypothetical statistical testing, specifically in null hypothesis significance testing. Statistical analysis is the best general way of examining data. These statistical tests are interpreted in combination with tables or graphs. There are two categories for statistical analysis, i.e., descriptive statistics and inferential statistics. Descriptive statistics provides a concise summary of data. It helps to describe/summarize data numerically or graphically in a meaningful way that includes things like standard deviation, variance, average, frequency, and range. Inferential statistics uses random data of samples (the data you collect) to describe inferences about a population. It becomes a valuable tool when it is not possible to draw general conclusions or to examine each member of an entire population. It is also used in predicting future of your experimental data [49].

Usually, the results obtained from different in vitro experiments cannot be transposed, for predicting any effect on an entire organism in vivo. For any scientific study, a consistent and reliable extrapolation of data from in vitro results to in vivo becomes extremely important. Thus it becomes imperative to increase the complexity of in vitro systems to obtain better understanding of in vitro results.

VALIDATION OF IN VITRO TOXICOLOGY TEST METHODS

On the basis of established in vivo-based toxicity assays, many new in vitro toxicity test methods have been developed. These newly developed toxicity tests have much importance in current scenario. Ames test has a great important in genotoxicity assessment of new compounds. Cell toxicity assays generally applied in the toxicological prescreening phase of the development of chemicals. In vitro cytotoxicity testing, biotransformation, and mechanistic investigations studies support the production of cost-effective new drugs. Newly generated in vitro toxicity test methods can easily reveal the antitumor activity of large numbers of compounds with accuracy. In regulatory testing situation these in vitro toxicity assays must be reliable, because from these tests it is only possible to generalize the data to incorporate into the risk assessment studies. The main purpose of validation studies is to distinguish methods, which are open to these postulates. Validation outcomes are beneficial for the updation of real protocols as well as for the extrapolation simulations and building of test batteries. Due to the difference in the accusatives of the different toxicity exercises, scientific provocation, and theoretical foundation, the principles for the substantiation of the test methods are not well concord. Information raised from the validation of toxicity exercises can be used for the improvement of the different in vitro toxicity tests.

In the year 1982, the first multicenter validation program was introduced by the Fund for the Replacement of Animals in Medical Experiments (FRAME) [50]. Some principles for the conducting interlaboratory validation schemes and design were published according to the FRAME program. For the validation process four different steps were distinguished: (1) interlaboratory validation, when the scientific quality of the new test protocol is canvassed; (2) during intralaboratory validation formulation of the method, its potencies limitations and mechanistic basis are determined; (3) regulatory validation, the process by which the test gets admitted by the regulatory authorities; and (4) extralaboratory validation, when the new test contends with other methods for usefulness in rendering data relevant to real-world troubles.

In 1990 Frazier fleshed out the idea of linearity, concept of predictive value and criteria for the choosing the tests and chemicals in the validation process. He further resolved that the intended use, cost, usability and scientific qualities must be reckoned with respect to the validation of the particular test [51]. According to one of the postulates, in the process of validation one of the main problems is to judge for a particular purpose how good a method is. The key question was whether this should be set up by the correlations of nonlinear or linear parameters, harmony, absolute correlations, or different criterion. Other decisive outcomes were whether in vitro performance should be compared with human or animal data and how we can select the reference chemicals and keep track of data generated.

For the validation of specific toxicological endpoints an international reference chemical data bank would be made available to research workers interested in validating alternative methods containing a collection of chemicals that have been carefully reviewed and selected; classified with respect to toxicity. The usefulness of the database can be enhanced if the results of validation studies were recorded in a central repository.

For the testing of new concepts in toxicology, validation studies in order to minimize ad-hoc explanations, the validation models must be predetermined before the actual validation takes place. Validation methods now include studies of the fit of data to a direct 1:1 correlation, linear regression with predetermined toxicity analysis of outliers, and then multivariate regression analysis [52].

THE "OMICS" APPROACH

The principle underlying the application of the "omics"—genomics, proteomics, and transcriptomics approach, for the evaluation of toxicity is based on the fact that the effect of a toxicant exposure on a biological system will almost without exception be reflected, at the cellular level on its gene expression. The changes in the expression level of different genes serve as measurable endpoints and the data generated can complement other established methodologies for accurate extrapolation of the in vitro results. Further, the measurement of mRNA and proteins can be useful in estimating the toxic potential of a chemical even before the development of a pathological response. Thus the advances in "omics" hold immense potential for its development into fast, reproducible, and reliable test systems.

The analysis of global gene expression changes helps in identifying diagnostic gene expressions which can be used to determine the toxic potential of agents and help in the development of new toxicity markers. The current use of DNA or protein chip platforms semiquantitatively measures the transcriptional activity of thousands of genes in a biological sample [53].

The information gathered from the genome based approach helps in correlating peptide data with gene sequences. Alterations in a particular set of proteins can help in identification of changes in a biochemical process that may relate to toxicity. High-throughput, automated proteomics techniques can be applied to analyze a range of biological samples like cell cultures, tissue samples, and body fluids for screening new markers of toxicity and exposure [54]. This approach will enable the toxicologists to study the effects of a single compound or of a mixture of chemicals on the entire gene expression and will speed up the screening of compounds for their toxicological properties and the selection of molecules for further development.

CONCLUSION

Cellular models that can explicitly recapitulate the in vivo physiologic and pathologic processes are the most advanced tool widely used for drug screening for toxicity assessments. Toxicity is a multifactorial organ-specific process often having a cell-specific etiology. Many processes that contribute to toxic insult require cellular communication with extracellular matrix (ECM) proteins. Therefore, an in vitro system such as immortalized cell lines grown on 2D surfaces may not be able to justify as a toxicological screen as it cannot discriminate these differences and lack cell ligands for ECM-cell adhesion molecule interaction. Development of more sophisticated 3D culture systems like tissue slices or hepatocyte sandwich culture preserves native ECM interactions and restores both chemical and mechanical tissue like stimulations. 3D assays that respond with organ-specific biomarkers can be useful in establishment of therapeutic window for determining the dose that organs might be exposed to in vivo.

Several other reasons hinder the development of in vitro systems as a model for toxicity evaluation. The changes in the biokinetics of the drug due to the presence of blood–brain barrier is not considered, whereas evaluation of neurotoxicity of the compound in neuronal system. Furthermore, the evaluation of toxicity of the parent compound in the target tissue would be irrelevant if its site of metabolism is different from the ultimate target of toxicity. Therefore, integrated testing strategies should be developed to overcome these limitations. Table 2.2 draws a comparison between the in vitro and in vivo test models by focussing on their pros and cons.

The advancement and implementation of in vitro models with appropriate resemblance to native tissue should be optimized. The tissue-specific information generated from these systems is accurate and reproducible and, therefore, improves the predictive value and translation efficiency between in vitro, animal and clinical studies.

ACKNOWLEDGMENTS

S.S. is thankful to the Indian Council of Medical Research for the award of Senior Research Fellowship. S.M. acknowledges the Department of Science and Technology (DST), Government of India, for providing financial assistance vide reference no. SR/WOS-A/LS-1290/2015 under Women Scientist Scheme. J.D. is thankful to University Grants Commission for the award of Senior Research Fellowship. A.D. is thankful to Department of Biotechnology for the award of Junior Research Fellowship. P.K.P. (PDF/2015/000033) is thankful to DST for providing financial assistance. This work was supported by Council of Scientific and Industrial Research Network project BSC0103. The CDRI communication no. for this manuscript is 9519.

TABLE 2.2 A Comparative Study of The Advantages and Disadvantages Associated With In Vivo and In Vitro Toxicity Assessment

In Vitro Toxicity Models	
Advantages	**Disadvantages**
• These are easy to set up and allow miniaturization as well as automation.	• Dissociation of the tissue leads to disruption of cellular structural integrity and impairs intracellular signalling.
• These are cost effective, yield quick results, permit replication and good quantification.	• There are differences between in vitro and in vivo compound biokinetics and the lack of biotransformation capabilities is probably the best-known limitation [68].
• Require little test substance.	• It is difficult to assess the concentration at which the drug is pharmacologically and toxicologically active and relevant to in vivo dose levels.
• Studies can be performed at the cellular or molecular level [65].	
• Cell models for practically all tissues or laboratory animal species are available [66].	• The cells are maintained in artificial non-physiological conditions as it does not reflect the body temperature of animals or the blood electrolyte concentrations of species [69].
• Monolayer cultures can be maintained for a long duration of time.	
• The effects of toxicants on functions requiring organization of cells can be studied using monolayer cultures [67].	• Culture conditions are not homeostatic, oxygen supply is not sufficient due to which anaerobic culture conditions are created indicated by phenol red turning yellow.
• There are few ethical problems, with the notable exceptions of human embryonic stem cells and human tissue donation.	• The immortalized cell lines used for these purpose accumulate numerous mutations including losses of parts or whole chromosomes [70].
• Novel approaches such as the image technologies as well as the diverse "omic" technologies will further improve the quality of data and reduce the time required to acquire it.	

In Vivo Models	
Advantages	**Disadvantages**
• These models exhibit extensive interactions among cells and tissues and mimic the physiological conditions similar to the human body.	• The inevitable killing of animals raises ethical concerns [72].
• Studies can be performed at the organ or system level.	• The human response cannot be predicted as the response to a toxicant varies amongst species [73] .
• The animal models contain the same molecular targets or pathways as humans [71].	• The use of inbred-strains does not reflect natural variances especially for uptake and biotransformation of test compound [74].
	• These models require relative large amounts of test substance.
	• Data are difficult to interpret because of the complexity of interactions.

REFERENCES

[1] Russell WMS, Burch RL, Hume CW. The principles of humane experimental technique. 1959.

[2] Hartung T, Balls M, Bardouille C, Blanck O, Coecke S, Gstraunthaler G, Lewis D. Good cell culture practice. Altern Lab Animals ATLA 2002;30:407–14.

[3] OECD. Test Guideline 491: reconstructed human cornea-like epithelium (RhCE) test method for identifying chemicals not requiring classification and labelling for eye irritation or serious eye damage. 2016.

[4] OECD. Test Guideline 442D: in vitro skin sensitisation: ARE-Nrf2 luciferase test method. 2015.

[5] OECD. Test Guideline 430: in vitro skin corrosion: transcutaneous electrical resistance test method (TER). 2013.

[6] OECD. Test Guideline 431: in vitro skin corrosion: reconstructed human epidermis (RHE) test method. 2016.

[7] OECD. Test Guideline 435: in vitro membrane barrier test method for skin corrosion. 2006.

[8] OECD. Test Guideline 439: in vitro skin irritation: reconstructed human epidermis test method. 2013.

[9] OECD. Test Guideline 428: skin absorption: in vitro method. 2004.

[10] OECD. Test Guideline 476: in vitro mammalian cell gene mutation tests using the Hprt and xprt genes. 2014.

[11] OECD. Test Guideline 492: in vitro mammalian cell gene mutation assays using the thymidine kinase gene. 2014.

[12] OECD. Test Guideline 487: in vitro mammalian cell micronucleus test. 2010.

[13] OECD. Test Guideline 473: in vitro mammalian chromosomal aberration test. 2013.

[14] Barile FA. Continuous cell lines as a model for drug toxicity assessment-3. 1997.

[15] Houser S, Borges L, Guzowski D, Barile F. Isolation and maintenance of continuous cultures of epithelial cells from chemically-injured adult rabbit lung. Altern Lab Animals ATLA 1990.

[16] Freshney RI. Basic principles of cell culture. Cult Cells Tissue Eng 2006;11:4.

[17] Briske-Anderson MJ, Finley JW, Newman SM. The influence of culture time and passage number on the morphological and physiological development of Caco-2 cells. Exp Biol Med 1997;214:248–57.

[18] Van der Valk J, Brunner D, De Smet K, Svenningsen ÅF Honegger P, Knudsen LE, Lindl T, Noraberg J, Price A, Scarino M. Optimization of chemically defined cell culture media – replacing fetal bovine serum in mammalian in vitro methods. Toxicol In Vitro 2010;24:1053–63.

[19] Patterson MK. Measurement of growth and viability of cells in culture. Methods Enzymol 1979;58:141–52.

[20] Zucco F, De Angelis I, Stammati A. Cellular models for in vitro toxicity testing. Animal cell culture techniques. Springer; 1998. p. 395–422.

[21] Astashkina A, Mann B, Grainger DW. A critical evaluation of in vitro cell culture models for high-throughput drug screening and toxicity. Pharmacol Ther 2012;134:82–106.

[22] Michel M, L'Heureux N, Pouliot R, Xu W, Auger FA, Germain L. Characterization of a new tissue-engineered human skin equivalent with hair. In Vitro Cell Dev Biol Animal 1999;35:318–26.

[23] Sundstrom L, Pringle A, Morrison B, Bradley M. Organotypic cultures as tools for functional screening in the CNS. Drug Discov Today 2005;10:993–1000.

[24] Allen DD, Caviedes R, Cárdenas AM, Shimahara T, Segura-Aguilar J, Caviedes PA. Cell lines as in vitro models for drug screening and toxicity studies. Drug Dev Ind Pharm 2005;31:757–68.

[25] Yang Z, Xiong H-R. In vitro, tissue-based models as a replacement for animal models in testing of drugs at the preclinical stages. INTECH Open Access Publisher; 2012.

[26] Pampaloni F, Reynaud EG, Stelzer EH. The third dimension bridges the gap between cell culture and live tissue. Nat Rev Mol Cell Biol 2007;8:839–45.

[27] Blaauboer BJ, Boekelheide K, Clewell HJ, Daneshian M, Dingemans MM, Goldberg AM, Heneweer M, Jaworska J, Kramer NI, Leist M. The use of biomarkers of toxicity for integrating in vitro hazard estimates into risk assessment for humans. Altex 2012;29:411–25.

[28] Eisenbrand G, Pool-Zobel B, Baker V, Balls M, Blaauboer B, Boobis A, Carere A, Kevekordes S, Lhuguenot J-C, Pieters R. Methods of in vitro toxicology. Food Chem Toxicol 2002;40:193–236.

[29] Anadon A, Martinez MA, Castellano V, Martinez-Larranaga MR. The role of in vitro methods as alternatives to animals in toxicity testing. Expert Opin Drug Metabol Toxicol 2014;10:67–79.

[30] Ames BN, McCann J, Yamasaki E. Methods for detecting carcinogens and mutagens with the Salmonella/mammalian-microsome mutagenicity test. Mutat Res Environ Mutagen Relat Subj 1975;31:347–63.

[31] Luepke N. Hen's egg chorioallantoic membrane test for irritation potential. Food Chem Toxicol 1985;23:287–91.

[32] OECD. Test No. 435: in vitro membrane barrier test method for skin corrosion. 2015.

[33] Spielmann H, Balls M, Dupuis J, Pape W, Pechovitch G, De Silva O, Holzhütter H-G, Clothier R, Desolle P, Gerberick F. The international EU/COLIPA in vitro phototoxicity validation study: results of phase II (blind trial). Part 1: the 3T3 NRU phototoxicity test. Toxicol In Vitro 1998;12:305–27.

[34] Mosmann T. Rapid colorimetric assay for cellular growth and survival: application to proliferation and cytotoxicity assays. J Immunol Methods 1983;65:55–63.

[35] Morgan DM. Tetrazolium (MTT) assay for cellular viability and activity. Polyam Protoc 1998:179–84.

[36] Slaughter M, Bugelski P, O'Brien P. Evaluation of Alamar Blue reduction for the in vitro assay of hepatocyte toxicity. Toxicol InVitro 1999;13:567–9.

[37] Gautheron P, Dukik M, Alix D, Sina JF. Bovine corneal opacity and permeability test: an in vitro assay of ocular irritancy. Toxicol Sci 1992;18:442–9.

[38] Collins FS, Gray GM, Bucher JR. Transforming environmental health protection. Science (New York, NY) 2008;319:906.

[39] Groten JP, Feron VJ, Sühnel J. Toxicology of simple and complex mixtures. Trends Pharmacol Sci 2001;22:316–22.

[40] Abramson SB, Weissmann G. The mechanisms of action of nonsteroidal antiinflammatory drugs. Arthritis Rheumatism 1989;32:1–9.

[41] Mitragotri S. Synergistic effect of enhancers for transdermal drug delivery. Pharm Res 2000;17:1354–9.

[42] Poloni V, Dogi C, Pereyra CM, Fernández Juri MG, Köhler P, Rosa CA, Dalcero AM, Cavaglieri LR. Potentiation of the effect of a commercial animal feed additive mixed with different probiotic yeast strains on the adsorption of aflatoxin B1. Food Addit Contam A 2015;32:970–6.

[43] Bulusu KC, Guha R, Mason DJ, Lewis RP, Muratov E, Motamedi YK, Cokol M, Bender A. Modelling of compound combination effects and applications to efficacy and toxicity: state-of-the-art, challenges and perspectives. Drug Discov Today 2016;21:225–38.

[44] Goldoni M, Tagliaferri S. Dose–response or dose–effect curves in in vitro experiments and their use to study combined effects of neurotoxicants. In Vitro Neurotoxicol Methods Protoc 2011:415–34.

[45] Frei E, Canellos GP. Dose: a critical factor in cancer chemotherapy. Am J Med 1980;69:585–94.

[46] Shoichet BK. Interpreting steep dose-response curves in early inhibitor discovery. J Med Chem 2006;49:7274–7.

[47] Barlow R, Bond SM, Bream E, Macfarlane L, McQueen D. Antagonist inhibition curves and the measurement of dissociation constants. Br J Pharmacol 1997;120:13–8.

[48] Singh V, Rana RK, Singhal R. Analysis of repeated measurement data in the clinical trials. J Ayur Integr Med 2013;4:77.

[49] Strand R, Fjelland R, Flatmark T. In vivo interpretation of in vitro effect studies with a detailed analysis of the method of in vitro transcription in isolated cell nuclei. Acta Biotheor 1996;44:1–21.

[50] Balls M, Horner S. The FRAME interlaboratory programme on in vitro cytotoxicology. Food Chem Toxicol 1985;23:209–13.

[51] Goldberg AM, Frazier JM, Brusick D, Dickens MS, Flint O, Gettings SD, Hill RN, Lipnick RL, Renskers KJ, Bradlaw JA. Framework for validation and implementation of in vitro toxicity tests. In Vitro Cell Dev Biol Animal 1993;29:688–92.

[52] Walum E, Clemedson C, Ekwall B. Principles for the validation of in vitro toxicology test methods. Toxicol In Vitro 1994;8:807–12.

[53] Nuwaysir EF, Bittner M, Trent J, Barrett JC, Afshari CA. Microarrays and toxicology: the advent of toxicogenomics. Mol Carcinogen 1999;24:153–9.

[54] Bach PH, Obatomi DK, Brant S, Castell J, Gomez-Lechon M. In vitro models for nephrotoxicity screening and risk assessment. In Vitro Methods Pharm Res 1996:55–101.

[55] Soldatow VY, LeCluyse EL, Griffith LG, Rusyn I. In vitro models for liver toxicity testing. Toxicol Res 2013;2:23–39.

[56] Harry GJ, Billingsley M, Bruinink A, Campbell IL, Classen W, Dorman DC, Galli C, Ray D, Smith RA, Tilson HA. In vitro techniques for the assessment of neurotoxicity. Environ Health Perspect 1998;106:131.

[57] Inoue T, Tanaka K, Mishima M, Watanabe K. Predictive in vitro cardiotoxicity and hepatotoxicity screening system using neonatal rat heart cells and rat hepatocytes. AATEX 2007;14:457–62.

[58] Aufderheide M, Knebel J, Ritter D. Novel approaches for studying pulmonary toxicity in vitro. Toxicol Lett 2003;140:205–11.

[59] Lankveld D, Van Loveren H, Baken K, Vandebriel R. In vitro testing for direct immunotoxicity: state of the art. Immunotox Test Methods Protoc 2010:401–23.

[60] Roguet R. Use of skin cell cultures for in vitro assessment of corrosion and cutaneous irritancy. Cell Biol Toxicol 1999;15:63–75.

[61] Spielmann H, Castle J, Gomez M. Ocular irritation. San Diego, CA, USA: Academic Press; 1997.

[62] Seiler AE, Spielmann H. The validated embryonic stem cell test to predict embryotoxicity in vitro. Nat Protoc 2011;6:961–78.

[63] Newall D, Beedles K. The stem-cell test – a novel in vitro assay for teratogenic potential. Toxicol In Vitro 1994;8:697–701.

[64] Li AP. Preclinical in vitro screening assays for drug-like properties. Drug Discov Today Tech 2005;2:179–85.

[65] Straughan DW, Fentem JH, Balls M. Replacement Alternative and Complementary In Vitro Methods in Pharmaceutical Research; 1997. p. 1.

[66] Davila JC, Rodriguez RJ, Melchert RB, Acosta Jr D. Predictive value of in vitro model systems in toxicology. Annu Rev Pharmacol Toxicol 1998;38(1):63–96.

[67] Maurel P. The use of adult human hepatocytes in primary culture and other in vitro systems to investigate drug metabolism in man. Adv Drug Deliv Rev 1996;22(1):105–32.

[68] Coecke S, Ahr H, Blaauboer BJ, Bremer S, Casati S, Castell J, Combes R, Corvi R, Crespi CL, Cunningham ML. Metabolism: a bottleneck in in vitro toxicological test development. The report and recommendations of ECVAM workshop 54. Altern Lab Animals ATLA 2006;34(1):49.

[69] Hartung T. Food for thought... on cell culture. Altex 2007;24(3):143.

[70] Ponten J. Cell biology of precancer. Eur J Cancer 2001;37:97–113.

[71] Hartung T, Daston G. Are in vitro tests suitable for regulatory use? Toxicol Sci 2009;111(2):233–7.

[72] Hartung T. Food for thought... on animal tests. Altex 2008;25(1):3–9.

[73] Lovell D. Impact of pharmacogenetics on toxicological studies. Statistical implications. J Exp Animal Sci 1993;35(5–6):259–81.

[74] Kacew S. Invited review: role of rat strain in the differential sensitivity to pharmaceutical agents and naturally occurring substances. J Toxicol Environ Health A 1996;47(1):1–30.

Chapter 3

Models and Methods for In Vitro Toxicity

Abhishek K. Jain, Divya Singh, Kavita Dubey, Renuka Maurya, Sandeep Mittal, Alok K. Pandey
CSIR-Indian Institute of Toxicology Research, Lucknow, India

INTRODUCTION

The term in vitro is derived from the Latin phrase, which means "the technique of performing a given procedure in an artificial environment outside the living organism". New tools such as in vitro models and computer-based assays are totally different from animal-based approaches as they provide a researcher a mechanistic overview of human relevant toxicological data. With an excellent study design and proper test system selection, these tools can predict some marked inferences.

To improve the scientific basis of current risk assessment procedures, there is a need for better understanding of the mechanism of chemical-induced toxicity. Thus the probable answer to this is the use of in vitro testing models, which provide a researcher with considerably more control over the variables and that it is relatively simple, inexpensive, and efficient [1]. Millions of chemicals are introduced every year in the global market whose toxic potential needs to be assessed [2]. It has been estimated that the cost of testing a single substance using whole animal is customarily above $2 million. The 3 Rs (3Rs) states replacement (with nonanimal models), reduction (of animal numbers), and refinement (to decrease animal suffering) is universally accepted on the basis of good laboratory practices [3,4]. A wide range of probing tools have developed that may possibly replace animal use within toxicity testing. In vitro methods are widely utilized for ranking of chemicals. Environmental Protection Agency has accepted a battery of in vitro test to assess large amount (volume) chemicals. A number of in vitro tests are gaining wide acceptance in order to replace in vivo cytogenetics with in vitro cytogenetics [5]. In recent years, Interagency Coordinating Committee on the Validation of Alternative Method and its European counterparts have validated a number of in vitro tests [6]. The most common ones generally assess the cell viability or toxicity.

In Vitro Toxicology. http://dx.doi.org/10.1016/B978-0-12-804667-8.00003-1

45

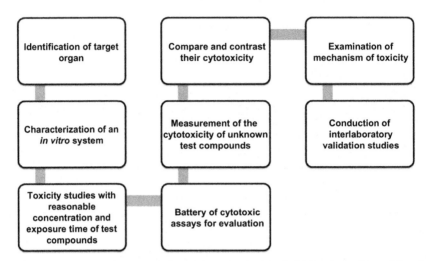

FIGURE 3.1 Systematic representation of phases involved in elucidating the in vitro toxicity of a test compound.

Although various in vitro test methods have been developed, only a few of them have been accepted or recommended by various governing bodies or régulatory agencies. Therefore, there is an urgent need to refine the existing in vitro methods or the new nonanimal model alternatives. The other thing is to increase the use of validated in vitro test models [7]. The optimum test protocol should have applicability in different laboratories and reproducibility in operational sense. In summary, it is established worldwide that in vitro testing models are of extreme importance for screening as well as to elaborate the mechanism and set of causes of various diseases (Fig. 3.1).

NEED OF IN VITRO MODELS FOR TOXICITY ASSESSMENT

Increase in ethical issues is the prime concern while using animal models for toxicity testing as it involves their inevitable killing [8]. Limited predictive capacity of animal test for determination of acute toxicity [9], reproductive toxicity [10] requirement of large amount of test substance, and most importantly use of inbred strains do not reflect natural variances, which suggests the use of the current prevailing scientific approach that reflects our mechanistic approach. The basic methodologies for testing these in vitro models are widely established. The framework or the set up involving in vitro test system requires little test substance, thereby reducing the overall cost.

The major aims of in vitro system were to develop, stimulate, and predict biological reactions to materials when placed into or on tissue in the body. In vitro biocompatibility tests were developed to study the original tissue characteristics and cell–cell interaction, creation of appropriate environment for cells, rapid performance of cytotoxicity assays, and sensitivity and accuracy in the identification of end-stage event [11].

VALIDATION OF IN VITRO TEST METHODS

Need of validation arises when there is a necessity to determine whether the new test is better than the existing animal test [7]. As a result various workshops were organized to discuss the concepts and practical aspects of validation [12,13]. Existing in vitro methods need to be revised and new test methods should be added regularly in order to improve the assessment of potential toxic chemicals on human health and environment. These in vitro test methods not only assist in hazard identification and risk assessment but also in understanding dose–response relationship and evaluating a new product before its market launch [14]. Examples of some proposed, accepted, and validated in vitro test methods for testing of different toxicity endpoints, which are prevalent because of use in safety testing of personal care, cosmetics, and chemicals in various industries are like genetic toxicology (OECD, 1997), phototoxicity (OECD, 2004d), and irritation (OECD, 2010) [15]. Any new or revised test method needs to be validated and accepted by regulatory agencies.

Criteria for Validation

The requirement of validation of test methods paves the future road for the success of the test. Therefore, the need arises to focus upon the prerequisites of obtaining proper validation:

- Rationale (scientific and regulatory) of the test method
- Relationship of biological effect and test method endpoint
- Detailed protocol along with the material methods and a brief overview of what is to be measured and how with a description of the limitation of the test.
- The validity of the test methods should be supported by various data in agreement with good laboratory practices.
- The test method performance, i.e., with test variability and reproducibility should be demonstrated using reference chemicals. Sufficient data should be provided as to allow for a comparison with the existing test and species of concern.

Criteria for Regulatory Acceptance

Validation alone does not make a method acceptable by the regulatory agencies, they are required to fit in the regulatory structure that has the following attributes:

- The method should be robust, efficient, transferable between labs, cost effective, and suitable for international acceptance.
- The method should have undergone independent scientific peer review by unbiased person who have gained expertise from different fields.

- The strength and weakness of the test methods should be clearly mentioned.
- There should be a detailed protocol with standard operating procedure and criteria for testing the performances and results.

Prior performing the validation study, the model being tested and the biological endpoint to predict the procedure should be clearly defined. Validation study should be planned in advance. To design and direct the validation study, framing of steering committee should be done. This committee would describe the aim of the study, select chemical, participating labs, and make sure that the test protocol is clearly defined and developed. After this, it would review and evaluate the data. The process for regulatory acceptance differs from one country to another as most of the government agencies accept methods that meet their respective need. Finally, international acceptance in the form of Organization for Economic Co-operation and Development (OECD) test guidelines is the most important milestone for validation.

MODELS FOR IN VITRO TOXICITY ASSESSMENT

In Vitro Models for Cytotoxicity Studies

Several in vitro models, such as monolayer, coculture, multilayer coculture, xenograft models, and tissue slice are especially developed to perform the cytotoxicity analysis of various chemicals. The cytotoxicity model system is useful in controlling cell adhesion, cell shape, intercellular contact, cell interactions, and provides remarkable benefits over convention in vitro systems. The common in vitro cytotoxicity models and their applications are listed in the table below (Table 3.1).

In Vitro Models for Specific Toxicity

Several in vitro models are established to assess toxic potential of chemicals. Herein are showed some models related to organ-specific toxicity of toxicants.

In Vitro Models for Liver Toxicity

Cell lines are extensively used for assessment of liver toxicity because they display similar genotypic and phenotypic characteristics of normal liver cells with functional enzymes responsible for phase I and phase II metabolism of xenobiotics [16,17]. Liver cell lines are the best choice for toxicological and pharmacological testing for detection of toxic chemicals and evaluating their cellular mechanism of toxicity. Liver toxicity implies chemical derived damage lead to several acute and chronic liver diseases. Many hepatic in vitro models are used in understanding of toxicity. Liver cell lines HepG2, Hep3B, HBG, and HepaRG are commonly immortalized liver derived cell lines used for in vitro assessment of liver toxicity study [18,19].

TABLE 3.1 Common In Vitro Cytotoxicity Models and Their Applications

In Vitro Models	Applications	References
Two-dimensional (2D) cell culture model	• Used to study basal function of cells. • Used to study number of parameter including vital solution, cytosolic enzyme release, cell growth and cloning efficiency, viability, and membrane integrity of cells. • Used to determine the maximum permissible dosage of new drugs.	[63]
2D coculture model	• To investigate heterotypic cell–cell interaction in a mixed in vitro coculture system that relates to bone regeneration.	[64] and [65]
Three-dimensional (3D) culture model	• Used to investigate the multiple feedback mechanism. • To investigate the role of adhesion molecule in invasion/metastasis.	[66]
Tissue slice/grafting models	• Used to study 3D representation of tissue slice, cell to cell and cell to matrix relationship. • Used to study of synergistic, additive or antagonistic effect of chemical.	[67]
Microscale cell culture model (microfluidic cell culture system)	• Used to study in vivo situation very closely. • Helps to study multiorgan interaction with blood circulation. • Along with 3D cell dimension used to study whole-body response to drug.	[68]

Adopted From Mahto SK, Chandra P, Rhee SW. In vitro models, endpoints and assessment methods for the measurement of cytotoxicity. Toxicol Environ Health Sci 2010;2(2):87–93.

In Vitro Models for Lung Toxicity

Different human-derived in vitro models have been used for assessment of lung toxicity studies [20,21]. Cell line is recognized as a useful in vitro model for the assessment of damaging effects induced by chemicals and has notably contributed to increase our knowledge about mechanism involved in pulmonary toxicity [22]. The A549 cell line has been extensively used in the study of the human lung damage caused by toxic substances such as nanoparticles (NPs) [23]. NPs are those molecules, which belong to nanoscale range that exerts its toxic effect by elevating intracellular the level of reactive oxygen species (ROS) [24]. Recent studies have identified an indirect genotoxic pathway involving inflammation as one of the mechanism underlying the carcinogenic effects of air pollution or diesel exhaust particles. Increase in the messenger RNA (mRNA) levels of proinflammatory cytokines and higher level of DNA strand breaks could be observed in A549 cultures exposed to these mixtures [25]. BEAS-2 B and 16HBE140 (human bronchial epithelial) cells are used in the field of pollutants, tobacco, and nanomaterials like in some cases these cell lines are being used to assess the immunological response to organic extracts isolated from air-borne particulate matter.

In Vitro Models for Neurotoxicity

In vitro systems are most successfully used to elucidate the mechanism of neurotoxicity and to describe the developmental changes induced by neurotoxicants [26]. Various cell lines are used to study the effect of various toxicants on neuronal cells such as neuroblastoma cells. Various classes of chemotherapeutic agents causing human peripheral neuropathies were identified and more than 30 chemicals that showed human neurotoxicants and neurite growth enhancer were correctly identified [27]. In vitro marker gene expression can be used to analyze the specification of neuronal and glial cell types as well as neuronal subtype (e.g., γ-aminobutyric acidergic neurons, glutaminergic neurons) [28].

In Vitro Models for Immunotoxicity

Humans are accidentally exposed to immunotoxic chemicals either occupationally or by contaminated food. The immune system plays a major role in maintaining human health, from the toxicological point of view, this system can be targeted from immunotoxic effects of variety of chemicals including the environmental pollutants like polychlorinated bisphenols, chlorinated dibenzo-p-doxins, pesticides and heavy metals, therapeutic drugs, and any other foreign substances often called as xenobiotics [29]. Therefore, exposure of such chemicals can result in immunosuppression that leads to decreased resistance to viral, bacteria, fungal, and other infectious agents. Heavy metals are considered to be immunosuppressive and ranked according to their immunosuppressive properties [30].

Mercury > Copper > Manganese > Cobalt > Cadmium > Chromium

Chemical-induced perturbation in the immune system led to immunosuppression, autoimmune disease, hypersensitivity reactions but no single immune test could be identified, which fully predicted for altered host defense. The apoptotic cell death assay was recently applied to lymphoid cell lines to study the effects of chemical immunotoxicant on apoptosis [31,32]. The majority of quantitative and functional assays were developed and validated for use in the quantification of cytokines, interferons, and other factors secreted by lymphocyte and macrophages cells [33]. Enzyme-linked immune sorbent assay as well as quantification of activated CD^{4+} and $CD8^+$ T-cell subset by flow cytometry clearly demonstrates chemical-induced dysregulation leading to autoimmune phenomena [34,35]. Human microglia (SV 40) and monocytic cell line (THP-1) are commonly used for immunotoxicity studies.

METHODS EMPLOYED FOR IN VITRO TOXICITY ASSESSMENT

Cytotoxicity Assessment

Cytotoxicity is the attribute of being toxic to the cells. Most of the assays having a broad range of different cytotoxic endpoints are now being used to measure the cellular responses toward a toxic chemical.

The choice for specific assay for analysis of cytotoxicity of a toxicant depends on various factors for instance, (1) in vitro cell culture models, (2) physicochemical properties of chemical, (3) culture platforms, (4) endpoints, (5) mechanism of cytotoxicity, and (6) assessment or detection methods. Various in vitro cytotoxicity assays have been utilized and classified on the basis of their endpoints. Nevertheless, selecting the right one from a variety of assays depend upon the need and experimental setup and design. The common cellular endpoints being measured for cytotoxicity analysis are listed in table below (Table 3.2):

MTT Assay

The MTT (3-(4, 5-dimethylthiazol-2-yl)-2,5-diphenyltetrazolium bromide) assay is a colorimetric assay for measuring cellular growth. It can be used for toxicity

TABLE 3.2 List of Cytotoxicity Assay Applied for Detection of Cellular Damage

Cell Parameters	Cytotoxicity Assay
Cell number	Trypan blue [69], methylene blue staining assay [69], ALP assay [70], resazurin [71], sulforodhamine B assay [62]
Cell viability	Lactate dehydrogenase (LDH) assay [62], Alamar blue [71], fluorescein diacetate [72], Calcein-AM [72]
Membrane permeability	3-(4, 5-Dimethylthiazol-2-yl)-2,5-diphenyltetrazolium bromide [62], LDH [62], annexin [72], granzyme-based [73], caspase-based assay [74]
Cellular adenosine triphosphate (ATP)	ATP-based luminescent assay [75]
Glucose	Fluorescent glucose analog [76]
Intracellular calcium	Fluo-322, Fluo-422 [77]
Lysosomal activity	Neutral Red assay [74], cathepsin D activity assay [78], granzyme-based assay [73]
Nuclear structure	Propidium dye [62], ethidium homodimer [62], BrdU [62], DAPI [62], TUNEL assay [74]
Total cellular protein	SRB assay [62]

assessment of toxicants. The water-soluble yellow dye MTT is a tetrazolium salt that is readily taken up by viable cells and reduced into purple color formazan by the action of mitochondrial succinate dehydrogenase in the mitochondria of living cells. Reduction of MTT occurs only in metabolically active cells, the level of activity is a measure of the viability of the cells. An organic solvent, usually dimethyl sulfoxide, is used to dissolve insoluble formazan crystal, a purple-colored product, which is measured by spectrophotometer. The amount of formazan produced is directly proportional to the number of viable cells present in the sample [36,37].

Neutral Red Dye Uptake Assay

The neutral red (NR; 3-amino-7dimethyl-2-methylphenazine hydrochloride) dye uptake is another cytotoxicity assay, which provides quantitative estimation of number of viable cells in a culture. It measures the ability of test compound to inhibit uptake of neutral red dye. NR is a weak cationic dye that penetrates into the cellular membranes and accumulates intracellularly in lysosomes. Viable cells incorporate NR dye into their lysosomes. As the cell surface alters or cell dies, their ability to uptake NR dye decreases. The absorbance is read using a spectrophotometer. Thus the loss of NR uptake inside lysosomes corresponds to loss of cell viability [38].

Propidium Iodide Uptake Assay

Flow cytometry provides a rapid and reliable method to quantify the live and dead cells in the cell suspension. This assay is dependent on the fluorescence intensity of the PI dye, which is intercalating between the DNA base pairs. PI dye uptake assay is commonly used for identifying dead cells in a given cell suspension. Generally, PI is not permeable to the intact cell membrane and excluded from viable cells. But in the case of dead cells, the cell membrane is damaged so the dye gets internalized inside the cells and binds the DNA and because of this its fluorescent intensity increases. The excitation and emission wavelength for PI dye is 488 and 617 nm.

Lactate Dehydrogenase Assay

Lactate dehydrogenase (LDH) is an oxidoreductase enzyme found in nearly all living cells (animals, plants, and prokaryotes) that is released into the cytoplasm upon cell lysis [39]. It is a colorimetric cytotoxicity assay that measures the membrane integrity. The level of LDH is more in damaged cells compared to normal cells. The LDH activity is measured on the basis of the conversion of lactate to pyruvate. LDH reduces nicotinamide adenine dinucleotide (NAD) to reduced NAD (NADH) and release H^+ ions; these ions catalyze reduction reaction of the tetrazolium salt (INT) to give the colored formazan compound, which shows the absorbance at 490–520 nm wavelength. The oxidation of NADH to NAD^+ is detected spectrophotometrically which show absorbance at 340 nm. NADH shows more absorbance in comparison to NAD^+ at 340 nm. The amount of color product formed is directly proportional to the activity of LDH in the sample.

Trypan Blue Exclusion Assay

Trypan blue exclusion assay was one of the most common and earliest method used for cell viability measurement [40]. It is impermeable for the normal cell membrane and therefore only enters the cell with compromised membrane. After entering the cell, it binds into the intracellular proteins and renders them bluish color. It also helps in the identification and counting of live or dead cells in the given cell population. The viable cells are shown small, rounded, and refractive, whereas dead cells are shown swollen, large, and dark blue.

Genotoxicity Assessment

In vitro test systems are known to determine the possible genotoxic potential of a test compound, which involves different stages of mutations: (1) gene and (2) chromosome. OECD has approved various test methods, which helps in determining genetic toxicology.

Comet (Single-Cell Gel Electrophoresis) Assay

This assay is a very attractive assay for the assessment of the DNA damage. Its simplicity, sensitivity, short time duration, and economy make it a prime choice in genotoxicity testing. Although this assay is described as a method for DNA damage quantification, the single-cell gel electrophoresis has evolved now to provide information about all kinds of DNA lesions. Comet assay is based on the supercoiled duplex DNA strand breakage. The comets are formed from the broken part of negatively charged DNA molecules and become free to move toward the anode when the electric field is applied. The formation of the comet is generally based on the migration of DNA and the number of broken ends and size of the DNA molecule.

Micronucleus Assay

Micronucleus (MN) is the extranuclear bodies of the damaged part of chromosome usually used to assess toxic potential of genotoxic agents. The study of DNA damage at the chromosome level is an essential part of genotoxicity testing because chromosomal mutation is an important event in carcinogenesis. This assay is gaining importance because it is used as an alternative to the chromosome aberration test. Initially, MN scoring was done with the help of Giemsa staining but later on cytokinesis-block MN method was developed for assessment of chromosomal genotoxic and mutagenic endpoints [41,42].

Chromosomal Aberration Assay

A chromosome is a strand of DNA, which transmitted genetic information from one generation to another. Chromosomal aberration (CA) is something that deviates from the normal way. CA usually occurs when cell division is disrupted. CA assay in cultured mammalian cells is a vital means for safety estimation of compounds, which have wide application in drugs, cosmetics, or food additives [43]. Genotoxic agents present in the environment induce heritable changes through chromosomal mutation, which can further lead to change in chromosome number and structure either directly by DNA strand break, base damage, hydrolysis of bases, etc. or indirectly by inhibition of DNA topoisomerases, generation of ROS, etc., which induces structural chromosomal changes. Different types of methods can be used to detect CAs such as microscopy, karyotyping, and fluorescence in situ hybridization.

Sister Chromatid Exchange Assay

Sister chromatid exchange (SCE) is the reciprocal exchange of chromatin between two identical sister chromatids. SCE possibly occurred during DNA synthesis either due to some replication error or due to inhibition of DNA replication [44]. SCE formation is an early indicator of chromosome instability in response to genotoxic agents. This assay examines the ability of a test chemical to increase the exchange of DNA in duplicating chromosomes between two sister chromatids. SCE is elevated in case of pathological conditions like Bloom syndrome (genetic disorder), Behcet disease (chronic inflammatory disorder), etc. SCE was first discovered by using Giemsa staining method. This method is able to stain in the presence of 5-bromodeoxyuridine (BrdU) base, which is introduced to the chromatin [45]. SCE formation is mediated by homologous recombination in vertebrate cells and the newly formed chromatid and their sisters are visualized cytologically, if one chromatid is labeled with 5-BrdU during synthesis [46].

Gamma-H2AX Assay

The H2AX, a variant of the histone family, contributes to the nucleosome formation and ultimately DNA structure. H2AX contains a unique SQ motif within its C-terminal tail that is highly conserved from plants to humans, suggesting a crucial role throughout evolution. The damage of DNA is a critical event able to affect cellular functions. Thus, it is essential for cells to maintain DNA integrity and repair such lesions effectively. Among different kinds of DNA lesions, double strand breaks (DSB) are considered to be the most critical type of DNA damage and misrepair can lead to tumorigenesis or cell death. A new biomarker, the phosphorylated histone H2AX, has become a powerful tool to monitor DNA DSBs. In response to DSBs, H2AX are rapidly phosphorylated on its serine residue by several kinases of phosphoinositol 3-kinases, specially ataxia telangiectasia mutated, then called γ-H2AX. γ-H2AX induction is one of the earliest events detected in cells following exposure to DNA damaging agents.

Cell Cycle and DNA Content Analysis

The quantitative analysis of cell cycle is a very crucial step to understanding the cellular mechanism and cell cycle progression as well as to assess the cell death [47]. In response to single and DSBs in DNA, certain check points arrest the cell cycle at certain stages. The cell cycle analysis and DNA content measurement of cells can be assessed by flow cytometry using nucleic acid staining dyes like propidium iodide (PI). PI enters the cells and binds to the DNA. However, the essential step is needed to exclude wrong interpretations. Such as, RNA could interfere in the staining. Therefore, solution must contain RNAase treatment before experiment has been performed. The principle is that the fluorescent dye (PI) binds with DNA in the cell suspension of permeabilized cells or nuclei. Fluorescence intensity of PI dye incorporated inside cells is directly proportional to the amount of DNA increase in cells [48]. However, it is well known that the presence of cells in sub-G1 phase of cell cycle correlates the DNA fragmentation with apoptosis [49].

Flow cytometry measurement offers the possibility to understand various parameters cumulatively, for example, cell cycle, apoptosis, and necrosis, which are useful tools for the screening toxicity effects of chemicals. It is reported previously that apoptotic cells in the liver play important role in the toxicity of xenobiotic metabolism. Tuschl and Schwab reported in their study that human liver cancer cell line (HepG2) is the best in vitro model for liver toxicity assessment and comparing the cell cycle effects of several chemical compounds such as acetaminophen (AAP), paraquat, malathion, 2,4-dichloro phenoxy acetic acid (2,4-D) and isoniazid (INH) on different cell lines such as human hepato carcinoma (HepG2), mouse lymphoma cell line (YAC-1), and human lymphoblastoid cell line (AHH-1) [49].

Apoptosis Detection Assay

The term "apoptosis", "active cell death", or "programmed cell death" is often used to describe a particular mode of cell death, which advances with specific change in the cytoplasmic and nuclear morphology of the cell [50]. Here, it is very important to note that programmed cell death can be of various types, such as apoptosis. Normally, apoptosis occurs during development, aging and can be a probable result of disease or poisonous substances.

Studies via light microscopy and transmission electron microscopy have revealed a number of morphological changes, which are considered to be the characteristic feature of apoptotic mode of cell death [51]. Apoptosis takes place via two separate stages; one is the formation of apoptotic bodies and second is their phagocytosis followed by degradation of cells. The process starts with nuclear shrinkage resulting in dense cytoplasm and tightly packed organelles. During the early stages of apoptosis, condensation of both nucleus and cytoplasm takes place followed by nuclear fragmentation and thereby the partition of protuberances that appear on the cell surface to produce many membranes bounded dense but otherwise bits and pieces of greatly varying size. In the beginning of the nuclear condensation, the electron dense nuclear chromatin aggregates toward the periphery in some cases can be uniformly distributed. Extensive plasma membrane blebbing takes place along with the formation of lucent cytoplasmic vacuoles and compact masses of nuclear material in some bodies with the organelle integrity still maintained enclosed in intact plasma membrane [52].

A large number of methods have been devised to detect apoptosis in cells. The methods should not just give the degree of cell death and also the respective pathway related to apoptotic cell death mode (Fig. 3.2).

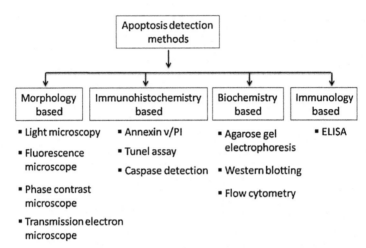

FIGURE 3.2 A schematic representation of different type of apoptosis detection methods.

Mutagenicity

To assess the mutagenic potential of any test compounds highly sensitive and specific test method is required. Mammalian gene mutagenesis test system in mammalian cells detects gene mutation at different gene loci. Several in vitro test methods are routinely used and accepted by regulatory authorities for assessment of mutagenicity.

Forward Gene Mutation Assay (Mammalian Model)

In vitro mammalian test system based on forward mutation, which measures the heritable changes in living cells caused by several possible mechanisms and capable of detecting gene mutation as well as genetic instability, which leads to cancer progression and development [53].

HGPRT Gene Mutation Assay

The Chinese hamster lung fibroblast (V-79) cell line is used to assess the mutagenic potential of chemical compounds inducing mutation at the heterozygous HGPRT gene locus, the gene that is present on the X-chromosome. This HGPRT gene encodes for the HGPRT enzyme, which is responsible for the conversion of the purines into nucleotides, toxic purine analog such as nucleobase guanine; for example, 6-thio guanine (6-TG) and 8-azaguanine (8-AG) are cytotoxic to cell having functional HGPRT enzyme (HGPRT+/−), whereas a forward mutation at the HGPRT gene locus (HGPRT−/−) induces resistance against toxic analog.

This assay is based on the metabolism of 6-TG and 8-AG by HGPRT to toxic derivatives that cause cell death or toxicity when incorporated into DNA synthesis during salvage biosynthesis. In this method, cells, for example, V-79 having HGPRT+/− are incubated with the test chemicals, replated in a medium containing 6-TG, mutations at the HGPRT gene locus inactivate the enzyme and inhibit the incorporation of toxic metabolites from 6-TG. Thus mutant cells proliferate and grow continuously using de novo purine biosynthesis and form colonies at a concentration of 6-TG normally cytotoxic to wild type cells [54,55].

Mouse Lymphoma Assay (L5178Y TK+/−)

The mouse lymphoma L5178Y mutagenesis system is based on quantifying mutations at the heterozygous thymidine kinase (TK+/−) gene locus. This TK gene encodes the TK enzyme, which is responsible for the incorporation of toxic analog of thymidine such as trifluorothymidine (TFT). When TFT is substituted for thymidine as the substrates in presence of TK enzymes, the resultant TFT inhibits thymidylate synthetase, resulting in cell death. Although mutation at the TK locus (TK−/−) results in the loss of TK activity induced resistance against toxic analog (TFT). When mouse lymphoma cells are exposed to mutagenic agent, forward mutation occurs at the TK locus and inactivates the activity of TK enzyme. Thus, TK deficient cells proliferate and form colonies using de novo biosynthesis, whereas wild type cells do not survive due to incorporation of toxic thymine analog [56] (Fig. 3.3).

Ames Test (Nonmammalian Model)

Ames test devised by a scientist "Bruce Ames" is used to assess the potential carcinogenic effect of chemicals by using the bacterial strain *Salmonella typhimurium*. This strain is mutant for the biosynthesis of histidine amino acid. As a result they are unable to grow and form colonies in a medium lacking histidine. When these mutant bacterial cells treated with chemicals, which are mutagenic causes a reversal of mutation in bacterial cells, which enables bacteria to grow on a media lacking in histidine [57]. More potency of a chemical leads to more number of cells forming colonies on Agar media. Chemicals found to be mutagenic in Ames test can be tested for their potential to induce carcinogenic

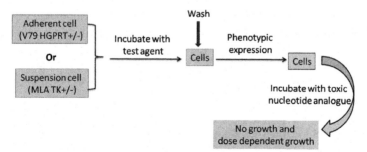

FIGURE 3.3 A hypothetical representation of forward gene mutation assay on either adherent cells (V79) or cells in suspension.

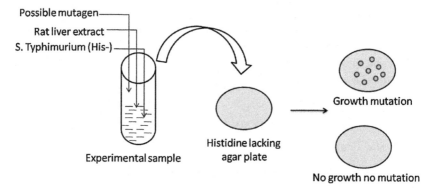

Possible mutagen
Rat liver extract
S. Typhimurium (His-)

Experimental sample

Histidine lacking
agar plate

Growth mutation

No growth no mutation

FIGURE 3.4 Principle of Ames test.

effect in animals. This test is commonly used by the pharmaceutical industry to test various drugs and chemicals before using them for clinical trials (Fig. 3.4).

Toxicokinetic Study

Toxicokinetic study is essentially required to relate the dose or chemical concentration and the mode of action of the chemicals and its various metabolites. The toxicokinetic process is responsible for the distribution and formation of various chemical entities at the target tissue, which is further responsible for determining the dose at the toxicological site. The rate and magnitude of absorption, distribution, metabolism, and excretion process determines the dose/concentration at the target tissue. The basic toxicokinetic parameter is based on in vitro and in silico studies, which detects the potential of accumulation and the potential of distribution or inhibition of chemicals in the tissues/organs. Toxicokinetic models can be divided into two broad categories depending on the function of time and dose: data-based compartmental models and physiologically based compartmental models.

In vitro approaches retrieve information of prime importance in the area of toxicokinetic studies. One of the representative model, the physiologically based toxicokinetic model, can be obtained by including the kinetics of metabolism by the liver and any other organ capable of biotransforming the compound (e.g., the lung), tissue–blood partition coefficients, and involving the transport process kinetics. Acute systemic toxicity of any chemical can be easily predicted by incorporating the basal cytotoxicity data along with the results revealed by toxicokinetic models, which purely depends on the concentration and time course in separate tissues.

Cellular and Functional Responses (Protein/Gene Expression)

The primary indicator of the effect of a toxic chemical on a biological system in most cases is its influence on gene expression, at the cellular level. Therefore, the measurement of the transcription (mRNA) and translation (protein) products of gene expression can provide important information about the potential toxicity of chemicals before the development of a toxic response. The rapid progress in genomic (DNA sequence), transcriptomics (gene expression), and proteomics (the study of proteins expressed by a genome, tissue or cell) technologies, in combination with the ever-increasing power of bioinformatics, creates a unique opportunity to form the basis of improved hazard identification for more predictive safety evaluation. Moreover, at present, the available methods for the study of gene expression at the transcript level include approaches based on hybridization, PCR, and sequencing-based techniques.

For screening purposes and detecting toxicity mechanism, two-dimensional gel electrophoresis is proving to be a highly sensitive technique, which combines separation of proteins by isoelectric focusing and sodium dodecyl sulfate-polyacrylamide gel electrophoresis based on molecular weight in the first dimension and the second dimension, respectively. When gene/proteins expressed after the exposure of a biological test system to a chemical are measured against those present under untreated conditions, it is possible to determine the changes in biochemical pathways via observed alteration in sets of gene/proteins that may be related to the toxicity, which provide the means to profile expression of thousands of messenger RNAs or proteins. During the past few years, the generation of ample research has resulted in an explosion of information regarding mechanism of toxicity and new tools to study the biological responses to toxic stress.

STATE OF ART AND RECENT DEVELOPMENT REGARDING IN VITRO APPROACHES

Omics Approach

Novel omics technology has a wide range of application. It involves an overall understanding of the molecules that makes up a cell, tissue, or organism. They are aimed fundamentally at the detection of genes (genomics), mRNA (transcriptomics), proteins (proteomics), and metabolites (metabolomics). These new fields are developing rapidly and now investigation is going on to integrate them with traditional testing techniques. These tools, techniques along with science provide a promising future in the advancement of test methods [58]. Genomics has revealed the current sequence of genes and proteins and now attention is on gene dynamics and interaction. Transcriptomics is the study of mRNA inside a cell or organism. The large-scale study of proteins along with their structure and function within a cell or organism is known as proteomics, whereas the study of global metabolite profile in a system is termed as metabolomics. The primary aim of using the -omics approach is the nontargeted identification of gene products, i.e., proteins, transcripts, and metabolites in a biological sample. Omics technology provides all the necessary tools required for understanding of the difference between DNA, RNA, proteins, and cellular molecules between different species and members of same species.

Bioinformatics and Computational Toxicology

Bioinformatics and databases of biological information can be used to create "maps" of cellular and physiological pathways and responses. Computational toxicology is a combination of mathematical and computer models to predict the response of any environmental agent and explain the series of events that follow on adverse effect. Bioinformatics and computational toxicology bridge the gap between data interpretation and software development. Its aim is to rapidly generate models for studying the functioning of cell, multicellular system, and finally the organism. It can generate virtual test systems for quick screening of toxic chemicals.

Integrated Testing Strategies

The design of testing strategies aims to make use of both existing and newly generated information to increase the quality of human safety assessment. Depending on the toxicological hazard assessed, there is a significant difference in testing strategies. There are various tests that stand alone for different parameters but a systematic combination of several information is often required. The need for involving different tests in integrated testing strategies (ITS) is as follows:

- Incorporating existing data
- Covering all the possible domains and mechanism
- Amelioration of the predictive value of the tests
- Addition of kinetic information to the test
- Filtering out costly and hectic animal tests because of less amount in test compounds

ITS can be described as an arrangement of test batteries covering important mechanistic steps and arranged in a hypothesis oriented form, which is of prime importance to make the efficient use of existing data so as to gain a summative understanding of the hazard or risk posed [59]. The first ITS was accepted as OECD guidelines in 2002 for eye and skin irritation (OECD TG 404,2002a; OECD TG 2002b) (Fig. 3.5).

CHALLENGES AND CONSIDERATIONS

In vitro test systems are extremely simplified form of very complex in vivo systems. Therefore, there is always a limitation with extrapolating the in vitro data with in vivo studies in toxicological situations. The biggest problem with in vitro systems is the lack of biotransformation studies [60]. The authenticity of in vitro cell lines is still a big issue as there is always a misconception with contamination and often the type of cell lines are mistaken [61]. As compared to the real world, the culture condition used in in vitro test systems are not homeostatic, i.e., depletion of nutrients, accumulation of waste

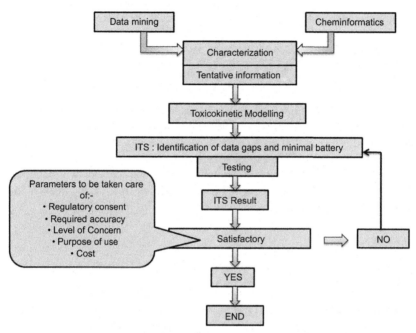

FIGURE 3.5 An outline of the protocol for opting optimal integrated testing strategies.

products, and limited dissolved oxygen supply, which sometimes results in anaerobic culture. As compared with the actual tissue situation, the cell density in in vitro culture is less than 1%, which disrupts intracellular signaling and the simulated nonphysiological conditions like relatively different body temperature as in animals, different blood electrolyte concentration, and the extracellular matrix.

Innovative, high-throughput in vitro screening methods should be developed to test various cell lines parallelly different chemicals and biological metabolites. Special consideration should be given to characterize in vitro cell-based assays so as to develop a possible understanding of the reaction in each cell of a particular assay.

FUTURE PERSPECTIVE

In the near future, work should be focused upon optimizing the existing in vitro assays and developing additional test to examine other pathways and target that could be important for toxicity assessment. It is anticipated that in the future, development, improvement, evaluation, and validation of test systems will be given an upper hand. Researchers should also expand the number of compounds to be tested to include nearly thousands of chemicals that humans are already exposed to that are in need of testing for potentially toxicological effects. The in vitro toxicity model is expected to have the characteristics of original tissue and its cell to cell interaction, conception of proper microenvironment for the cells, rapid performance, sensitivity, and accuracy.

CONCLUSION

It is expected that continued research in these models will contribute our limited understanding of the mechanism of toxicity as well as enhance their predictive abilities and utility in risk assessment. This study may be potentially useful for clarifying accorded mechanism of toxicity in human cell. Such efforts combined with many established toxicology principles and methods will definitely increase the degree of confidence in the extrapolation from developmental animals to human.

REFERENCES

[1] Garle M, Fentem J, Fry J. In vitro cytotoxicity tests for the prediction of acute toxicity in vivo. Toxicol In Vitro 1994;8(6):1303–12.

[2] Kniewald J, et al. Alternative models for toxicity testing of xenobiotics. Arhiv za Higijenu Rada I Toksikologiju (Archiv Ind Hyg Toxicol 2005;56(2):195–204.

[3] Knight A. Non-animal methodologies within biomedical research and toxicity testing. ALTEX 2008;25(3).

[4] Russell W, Burch RL. The principles of humane experimental technique. 1959. 1912.

[5] Walum E, Clemedson C, Ekwall B. Principles for the validation of in vitro toxicology test methods. Toxicol In Vitro 1994;8(4):807–12.

[6] Council NR. Toxicity testing for assessment of environmental agents: interim report. Washington, DC: National Academies Press; 2006.

[7] Frazier JM. Validation of in vitro models. Int J Toxicol 1990;9(3):355–9.

[8] Hartung T. Food for thought... on alternative methods for cosmetics safety testing. ALTEX 2008;25(3):147–62.

[9] Ekwall B, et al. MEIC evaluation of acute systemic toxicity: part V. Rodent and human toxicity data for the 50 reference chemicals. Altern Lab Animals ATLA 1998;26:571–616.

[10] Bremer S, et al. The development of new concepts for assessing reproductive toxicity applicable to large scale toxicological programmes. Curr Pharmaceut Design 2007;13(29):3047–58.

[11] Zucco F, et al. Toxicology investigations with cell culture systems: 20 years after. Toxicol In Vitro 2004;18(2):153–63.

[12] Balls M. Report and recommendations of the CAAT/ERGATT workshop on the validation of toxicity test procedures. Atla 1990;18:313–37.

[13] Eaton J, Kortum S. Trade in ideas Patenting and productivity in the OECD. J International Econ 1996;40(3):251–78.

[14] Andersen ME, Krewski D. Toxicity testing in the 21st century: bringing the vision to life. Toxicol Sci 2009;107(2):324–30.

[15] Costin GE, Raabe H. In vitro toxicology models. Role Study Dir Nonclin Stud 2014:145–70.

[16] Scheers E, Ekwall B, Dierickx P. In vitro long-term cytotoxicity testing of 27 MEIC chemicals on Hep G2 cells and comparison with acute human toxicity data. Toxicol In Vitro 2001;15(2):153–61.

[17] Sassa S, et al. Drug metabolism by the human hepatoma cell, Hep G2. Biochem Biophys Res Commun 1987;143(1):52–7.

[18] Guguen-Guillouzo C, Corlu A, Guillouzo A. Stem cell-derived hepatocytes and their use in toxicology. Toxicology 2010;270(1):3–9.

[19] Guguen-Guillouzo C, Guillouzo A. General review on in vitro hepatocyte models and their applications. Hepatocytes Methods Protoc 2010:1–40.

[20] Nichols WK, et al. 3-methylindole-induced toxicity to human bronchial epithelial cell lines. Toxicol Sci 2003;71(2):229–36.

[21] Van Vleet TR, Macé K, Coulombe RA. Comparative aflatoxin B1 activation and cytotoxicity in human bronchial cells expressing cytochromes P450 1A2 and 3A4. Cancer Res 2002;62(1):105–12.

[22] Castell JV, Donato MT, Gómez-Lechón MJ. Metabolism and bioactivation of toxicants in the lung. The in vitro cellular approach. Exp Toxicol Pathol 2005;57:189–204.

[23] Choi S-J, Oh J-M, Choy J-H. Toxicological effects of inorganic nanoparticles on human lung cancer A549 cells. J Inorg Biochem 2009;103(3):463–71.

[24] Nel A, et al. Toxic potential of materials at the nanolevel. Science 2006;311(5761):622–7.

[25] Dybdahl M, et al. Inflammatory and genotoxic effects of diesel particles in vitro and in vivo. Mutat Res Gen Toxicol Environ Mutagen 2004;562(1):119–31.

[26] Harry GJ, et al. In vitro techniques for the assessment of neurotoxicity. Environ Health Perspect 1998;106(Suppl. 1):131.

[27] Hoelting L, et al. Stem cell-derived immature human dorsal root ganglia Neurons to identify peripheral neurotoxicants. Stem Cells Translat Med 2016. p. sctm. 2015-0108.

[28] Brosamle C, Halpern ME. Characterization of myelination in the developing zebrafish. Glia 2002;39(1):47–57.

[29] Krzystyniak K, Tryphonas H, Fournier M. Approaches to the evaluation of chemical-induced immunotoxicity. Environ Health Perspect 1995;103(Suppl. 9):17.

[30] Lawrence D. Immunotoxicity of heavy metals. In: Dean JH, Luster ML, Munson AE, Amos H, editors. Immunotoxicology and immunopharmacologya. 1985.

[31] Burchiel SW, et al. DMBA induces programmed cell death (apoptosis) in the A20. 1 murine B cell lymphoma. Toxicol Sci 1993;21(1):120–4.

[32] El Azzouzi B, et al. Cadmium induced apoptosis in a human T cell line. Toxicology 1994;88(1):127–39.

[33] Luster MI, et al. Risk assessment in immunotoxicology II. Relationships between immune and host resistance tests. Toxicol Sci 1993;21(1):71–82.

[34] Gleichmann E, et al. Testing the sensitization of T cells to chemicals. From murine graft-versus-host (GvH) reactions to chemical-induced GvH-like immunological diseases. Autoimmunity and toxicologyAmsterdam: Elsevier; 1989. p. 363.

[35] Krzystyniak K, et al. Activation of CD4+ and CD8+ lymphocyte subsets by streptozotocin in murine popliteal lymph node (PLN) test. J Autoimmunity 1992;5(2):183–97.

[36] Mosmann T. Rapid colorimetric assay for cellular growth and survival: application to proliferation and cytotoxicity assays. J Immunol Methods 1983;65(1–2):55–63.

[37] Morgan DML. Tetrazolium (MTT) assay for cellular viability and activity. In: Morgan DML, editor. Polyamine protocols. Totowa, NJ: Humana Press; 1998. p. 179–84.

[38] Repetto G, del Peso A, Zurita JL. Neutral red uptake assay for the estimation of cell viability/cytotoxicity. Nat Protoc 2008;3(7):1125–31.

[39] Wroblewski F, Ladue JS. Serum glutamic pyruvic transaminase in cardiac and hepatic disease. Exp Biol Med 1956;91(4):569–71.

[40] Strober W. Trypan blue exclusion test of cell viability. Curr Protoc Immunol 2001. p. A3. B. 1–A3. B. 3.

[41] Fenech M, Morley AA. Measurement of micronuclei in lymphocytes. Mutat Res Environ Mutagen Relat Subj 1985;147(1–2):29–36.

[42] Luzhna L, Kathiria P, Kovalchuk O. Micronuclei in genotoxicity assessment: from genetics to epigenetics and beyond. Front Genetics 2013;4:131.

[43] Galloway SM. Chromosome aberrations induced in vitro: mechanisms, delayed expression, and intriguing questions. Environ Mol Mutagen 1994;23(S2):44–53.

[44] Morales-Ramírez P, Rodríguez-Reyes R, Vallarino-Kelly T. Fate of DNA lesions that elicit sister-chromatid exchanges. Mutat Res Fundamental Mol Mech Mutagen 1990;232(1):77–88.

[45] Stults DM, Killen MW, Pierce AJ. The sister chromatid exchange (SCE) assay. Mol Toxicol Protoc 2014:439–55.

[46] Sonoda E, et al. Sister chromatid exchanges are mediated by homologous recombination in vertebrate cells. Mol Cell Biol 1999;19(7):5166–9.

[47] Tao D, et al. New method for the analysis of cell cycle–specific apoptosis. Cytom A 2004;57(2):70–4.

[48] Nunez R. DNA measurement and cell cycle analysis by flow cytometry. Curr Issues Mol Biol 2001;3:67–70.

[49] Tuschl H, Schwab CE. Flow cytometric methods used as screening tests for basal toxicity of chemicals. Toxicol In Vitro 2004;18(4):483–91.

[50] Duvall E, Wyllie A, Morris R. Macrophage recognition of cells undergoing programmed cell death (apoptosis). Immunology 1985;56(2):351.

[51] Darzynkiewicz Z, Li X, Gong J. Assays of cell viability: discrimination of cells dying by apoptosis. Methods Cell Biol 1994;41:15–38.

[52] Kerr JF, Wyllie AH, Currie AR. Apoptosis: a basic biological phenomenon with wide-ranging implications in tissue kinetics. Br J Cancer 1972;26(4):239.

[53] Eastmond DA, et al. Mutagenicity testing for chemical risk assessment: update of the WHO/IPCS Harmonized Scheme. Mutagenesis 2009;24(4):341–9.

[54] Co-operation, O.f.E. and Development. Test No. 476: in vitro mammalian cell gene mutation test. OECD Publishing; 1997.

[55] Doak S, et al. In vitro genotoxicity testing strategy for nanomaterials and the adaptation of current OECD guidelines. Mutat Res Genetic Toxicol Environ Mutagen 2012;745(1):104–11.

[56] Clive D, et al. A mutational assay system using the thymidine kinase locus in mouse lymphoma cells. Mutat Res Fundamental Mol Mech Mutagen 1972;16(1):77–87.

[57] ZhongHua Z, et al. Assessment of the potential mutagenicity of organochlorine pesticides (OCPs) in contaminated sediments from Taihu Lake, China. Mutat Res Genet Toxicol Environ Mutagen 2010;696(1):62–8.

[58] Debnath M, Prasad GB, Bisen PS. Molecular diagnostics: promises and possibilities. Springer Science & Business Media; 2010.

[59] Jaworska J, Hoffmann S. Integrated testing Strategy (ItS)–Opportunities to better use existing data and guide future testing in toxicology. Altex 2010;27(4):231–42.

[60] Coecke S, et al. Metabolism: a bottleneck in in vitro toxicological test development. The report and recommendations of ECVAM workshop 54. Altern Laboratory Animals ATLA 2006;34(1):49.

[61] Buehring GC, Eby EA, Eby MJ. Cell line cross-contamination: how aware are Mammalian cell culturists of the problem and how to monitor it? In Vitro Cell Dev Biol Animal 2004;40(7):211–5.

[62] Mahto SK, Chandra P, Rhee SW. In vitro models, endpoints and assessment methods for the measurement of cytotoxicity. Toxicol Environ Health Sci 2010;2(2):87–93.

[63] Bourdeau P, et al. Short-term toxicity tests for non-genotoxic effects. 1990.

[64] Burguera EF, Bitar M, Bruinink A. Novel in vitro co-culture methodology to investigate heterotypic cell-cell interactions. Eur Cell Mater 2010;19(166):e79.

[65] Lu HH, Wang I-NE. Multiscale coculture models for orthopedic interface tissue engineering. Biomed Nanostruc 2007:357.

[66] Kim JB, Stein R, O'Hare MJ. Three-dimensional in vitro tissue culture models of breast cancer—a review. Breast Cancer Res Treatment 2004;85(3):281–91.

[67] Lipman J, et al. Cell culture systems and in vitro toxicity testing. Cytotechnology 1992;8(2):129–76.

[68] Sung JH, Shuler ML. In vitro microscale systems for systematic drug toxicity study. Bioproc Biosyst Eng 2010;33(1):5–19.

[69] Oliver MH, et al. A rapid and convenient assay for counting cells cultured in microwell plates: application for assessment of growth factors. J Cell Sci 1989;92(3):513–8.

[70] Huschtscha LI, Lucibello FC, Bodmer WF. A rapid micro method for counting cells "in situ" using a fluorogenic alkaline phosphatase enzyme assay. In Vitro Cell Dev Biol 1989;25(1):105–8.

[71] Al-Nasiry S, et al. The use of Alamar Blue assay for quantitative analysis of viability, migration and invasion of choriocarcinoma cells. Hum Reproduction 2007;22(5):1304–9.

[72] Lecoeur H, et al. A novel flow cytometric assay for quantitation and multiparametric characterization of cell-mediated cytotoxicity. J Immunol Methods 2001;253(1):177–87.

[73] Bird CH, et al. Cationic sites on granzyme B contribute to cytotoxicity by promoting its uptake into target cells. Mol Cell Biol 2005;25(17):7854–67.

[74] Reddivari L, et al. Anthocyanin fraction from potato extracts is cytotoxic to prostate cancer cells through activation of caspase-dependent and caspase-independent pathways. Carcinogenesis 2007;28(10):2227–35.

[75] Castano A, Tarazona J. ATP assay on cell monolayers as an index of cytotoxicity. Bull Environ Contamination Toxicol 1994;53(2):309–16.

[76] Leira F, et al. Fluorescent microplate cell assay to measure uptake and metabolism of glucose in normal human lung fibroblasts. Toxicol In Vitro 2002;16(3):267–73.

[77] Venkatesh N, et al. Chemical genetics to identify NFAT inhibitors: potential of targeting calcium mobilization in immunosuppression. Proc Natl Acad Sci USA 2004;101(24):8969–74.

[78] Lee DC, et al. 6-Hydroxydopamine induces cystatin C-mediated cysteine protease suppression and cathepsin D activation. Neurochem Int 2007;50(4):607–18.

Chapter 4

In Vitro Gene Genotoxicity Test Methods

Manisha Dixit, Amit Kumar

CSIR-Indian Institute of Toxicology Research, Lucknow, India

INTRODUCTION

Genetic materials form the basis of every cell and allow them to pass on the information from one generation to another. Any insult incurred by the cell leads to altered genetic material and thus, damage to the intricate cell machinery and finally, cell death.

Humans are constantly exposed to these insults due to exposure to ultraviolet rays or carcinogens like tobacco smoke, food adulterants, and so on. Following the industrial revolution, there was an indiscriminate use of pesticides and other effluents, which continued to accumulate in the environment with the advent of time [1]. Humans are thus unknowingly exposed to such compounds, which are deleterious to their health. These compounds become integrated into the DNA and cause mutations and thus, DNA damage. Any chemical capable of rendering DNA damaged is, therefore, termed genotoxic [2]. Over the past few decades, there have been immense advances in toxicological studies, which help us to identify various genotoxic agents. Whenever a cell is exposed to such genotoxic agents, the changes it introduces in the DNA can lead to carcinogenicity or pathogenesis of certain diseases like atherosclerosis or cardiac dysfunction [3]. The various types of DNA damage are explained in Fig. 4.1.

Genotoxicity is thus defined as the harmful, deleterious effect of any chemical that could potentially damage the cell's own genetic material, its DNA or RNA, thus affecting its integrity and normal function [2]. These genotoxins are often mutagens and include both chemical genotoxins and radiation. Genotoxins can be carcinogens, mutagens, or teratogens. Any substance that is genotoxic can damage the cell via interaction with the DNA sequence or with the structure and in the worst case can lead to manifestations of the gravest of serious diseases. As a consequence, DNA damage occurs in the form of strand breaks, single or double; adduct formation, or cross-linking of the bases, which are incorporated as errors

In Vitro Toxicology. https://doi.org/10.1016/B978-0-12-804667-8.00004-3

67

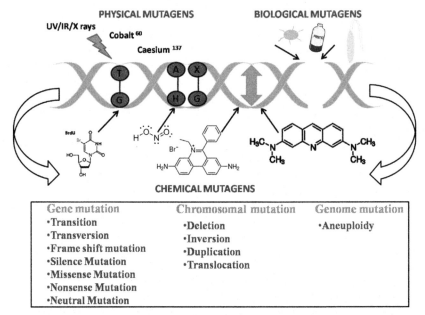

FIGURE 4.1 **Different kinds of mutagens and the types of mutation.** Different types of physical, chemical, and biological mutagens and the types of damages that they inflict upon the DNA. Lower panel: Different types of mutation caused by the various mutagens.

during DNA replication. These errors exacerbate during replication and can cause cell cycle arrest and ultimately, a faulty phenotype or cell death.

Genotoxicity is often confused with mutagenicity, but it is important to note that all mutagens are genotoxic agents, but all genotoxic substances need not be mutagens; for example, o-anthranilic acid, eugenol, and D,L-menthol are genotoxic substances with nonmutagenic activity [4]. Any agent that causes damage to chromosomes or DNA is referred as genotoxin. The hereditary changes caused by these genotoxic agents can affect somatic cells, resulting in somatic mutations such as cancer, or can be incorporated in the germ cell as a germ-line mutation and passed on as a heritably altered trait.

IMPORTANCE OF GENOTOXICITY TESTING

With the increase in advances in science, new drugs are tested every day to cure diseases. As a part of the safety evaluation process, the genotoxic potential of each drug must be checked and validated by the regulatory authorities. Any new chemical entity (NCE) introduced as a potential drug must be tested for a basic toxicological profile. The data thus obtained are used for efficacy and safety of the NCE based on risks versus benefit associated with the potential drug, and it is assessed by several guidelines such as Organization for Economic Cooperation and Development (OECD) or International Conference on Harmonization (ICH) guidelines. Genotoxicity testing, therefore, becomes pivotal; it is prerequisite for any new drug and is a mandatory protocol established by the regulatory authorities. Most countries have their own set of guidelines through which a new drug must pass before entering the market. Table 4.1 enlists the different regulatory guidelines set by various countries.

Despite all these protocols, the guidelines set by the OECD are crucial for attaining acceptance of toxicity data from OECD member countries. Similarly, there are the ICH guidelines, a combination of Europe, Japan, and US guidelines. Both OECD and ICH guidelines recommend the same test for the majority of chemicals and, by complying with either of these two standards, the toxicity data can be accepted internationally.

The genotoxicity tests ensure that the potential drug or NCE is not a potent carcinogen, mutagen, or teratogen.

TABLE 4.1 The Regulatory Guidelines of Different Countries

S.No.	Issued by	Guidelines
1.	Ministry of Health and Family Welfare, India	Drugs and Cosmetic Rules 1988 Central Drugs Standard Control Organization (CDSCO) Directorate General of Health Services (DGHS)
2.	Centre for Drugs and Biologics Evaluation and Research (CDER and CBER) Food and Drug Administration (FDA) USA	Food and Drugs—Title 21
3.	Japan Ministry of Health and Welfare (MHW), 1990; 1992 Ministry of Agriculture, Forestry, and Fisheries (MAFF,1985); Ministry of Labor (1991)	Japanese Guidelines for Mutagenicity Testing
4.	Health Protection Branch, Canada	Food and Drugs Act (R.S.C., 1985, c. F-27)
5.	Japan, USA, Europe	ICH guidelines
6.	International Conference on Harmonization (OECD) member countries	OECD guidelines
7.	European Union	91/356/EEC, 2003/94/EC and 91/412/EEC

TEST FOR GENOTOXICITY

Genotoxicity testing helps to identify the culprit responsible for somatic or germ cell alteration. Unlike other toxicity, genotoxicity is hazardous in the long term, as the insult might be present at a very low dose, but continuous exposure to such a subtle dose can also culminate in damaged DNA. To test whether a specific compound might be a genotoxic agent, the European Union has adopted in vitro genotoxicity test methods under the OECD guidelines. These guidelines are amended routinely according to current scientific advances and to improve regulatory needs and animal considerations [5]. For example, an assay that used the thymidine kinase (TK) locus was originally a part of Test Guideline (TG) 476, adopted in 1984, but now is a part of TG 490 due to certain amendments made by OECD members. Similarly, if certain tests are found to be irrelevant or no longer required, they are removed from the OECD listed guidelines. The various in vitro genotoxic tests are listed in Table 4.2.

At its 22nd meeting in March 2010, the Working Group of National Coordinators of the OECD initiated consultations to make amendments in the existing eight test methods and to update them. Subsequently, in 2013, TGs 477, 479, 480, 481, 482, 484 were annulled officially and the remainder were subjected to revision [5]. Though the deleted tests can no longer be used as standard testing protocols and should not be recommended as a part of OECD testing guidelines, the data continue to be helpful for risk assessment.

Of the OECD prescribed genotoxic tests, four in vitro assays are commonly used for analysis. Of these, one mutagenicity test method is based on bacterial cells and the others are carried out in mammalian cells. The bacterial cell-based tests include bacterial reverse mutation test (Ames test; TG 471), whereas the mammalian cell-based tests are the in vitro mammalian chromosome aberration test (TG 473) and the in vitro mammalian cell gene mutation test (OECD TG 476), the mammalian cell micronucleus test (TG 487), and mammalian cell gene mutation tests using the TK gene (TG 490).

THE STANDARD TEST BATTERY FOR GENOTOXICITY

To test the efficacy of any new drug, three levels of information (gene, chromosome and genome) are required before it can be registered as a potential therapeutic. The genotoxic tests, in combination with pharmacokinetic and toxicokinetic data, help to correlate toxic signs and symptoms with the blood and tissue levels of the agent and help in selecting suitable animal species. A battery of in vitro and in vivo tests are, therefore, chosen for the toxicological evaluation process.

The outline of a general battery test approach includes the following:

- **A test for gene mutation in bacteria.** Normally, the bacterial reverse mutation test or the Ames test is used, which identifies the majority of genotoxic rodent carcinogens.

TABLE 4.2 Current Status of the In Vitro Test Guidelines for Genetic Toxicology

S.No.	Test	Adopted	Revised	Deleted
TG 471	Bacterial reverse mutation test (Ames test)	1983	1997	
TG 472	*Escherichia coli*, reverse assay	1983		1997
TG473	Mammalian chromosome aberration test	1983	1997/2014	
TG476	Mammalian cell gene mutation test	1984	1997/2014	
TG479	Sister chromatid exchange assay in mammalian cells	1986		2013
TG482	DNA damage and repair, unscheduled DNA synthesis in mammalian cells	1986		2013
TG487	Mammalian cell micronucleus test	2010	2014	
TG490	Mammalian cell gene mutation tests using the thymidine kinase gene	2015		

- **An in vitro test with cytogenetic evaluation of chromosomal damage with mammalian cells or an in vitro mouse lymphoma TK assay:** DNA damage is tested in mammalian rather than bacterial cells for gross chromosomal damage, which includes structural and numerical chromosomal aberration tests.
  - Primary gene mutations are verified by the *tk* locus using either the L5178Y cells (mouse lymphoma) or human lymphoblastoid TK6 cells; the *hprt* locus using different cell lines like V79 cells, CHO cells, or L5178Y cells; and the *gpt* locus using AS52 cells.
  - Gene mutation and clastogenic mutations are identified by a mouse lymphoma *tk* assay that includes a wide range of mutations such as point mutation, deletions, insertions, translocations, etc.
- **An in vivo test for chromosomal damage using rodent hematopoietic cells:** The in vitro tests are followed by an in vivo test that provides a test

model in which additional factors are taken into account that could also affect the activity of a genotoxic compound. These include absorption, distribution, metabolism, and excretion studies, which help to detect some additional genotoxic agents. Thus, in vivo rodent hematopoietic cells are used that fulfill all these criteria.

The compounds under study are tested via the standard battery of test. If they show uniform negative results for all four tests performed in accordance with the current OECD guidelines, they are said to be nontoxic. Further tests are required in case of nonuniform results and positive results (ICH guidelines [5a]).

GENERAL PROTOCOL FOR IN VITRO TESTS

Though different in vitro tests have different test methods, a general protocol that every test follows is outlined below:

1. **Metabolic Activation:** Bacteria or the cells before exposure to the test substance must be activated in order to function as a potential mutagen. A commonly used system involves a cofactor-supplemented postmitochondrial fraction (S9) prepared from rodent livers. The rat livers are then treated with agents such as Aroclor or a combination compounds like phenobarbitone and β-naphthoflavone to allow enzymatic induction [6–8].
2. **Test substance/preparation**: If solid, the test substance is dissolved in appropriate solvent. Liquid substances or dissolved solids are often diluted if necessary.
3. **Solvent/vehicle**: The solvent/vehicle used should not react with the test compound and should not have deleterious effects either for the cell/bacteria or for the metabolic activator.
4. **Exposure concentrations**: There are recommended values for each test compound to be classified as cytotoxic or noncytotoxic. A maximum of 5 mg/plate or 5 μL/plate is acceptable for noncytotoxic compounds; although for compounds that are cytotoxic at lower levels, the cytotoxic concentrations for those should be validated. Five different concentrations should be used for each test substance.
5. **Controls**: Each assay should include a strain-specific positive and negative control, both in the presence and absence of the metabolic activator. A list of these is given in Table 4.3.
6. **Media and culture conditions:** Each cell line chosen for different assays is grown in its own medium, followed by correct incubation condition in humidified vessel at 5% CO_2 and 37°C. The cell lines should be routinely checked for modal number and contamination should be avoided.

TABLE 4.3 Different Types of Positive Control for Different In Vitro Test

Test	Positive Control (With Metabolic Activator)	Positive Control (Without Metabolic Activator)
Ames test	• 9,10-Dimethylanthracene • 7,12-Dimethylbenzanthracene • Congo Red • Benzo(a)pyrene • Cyclophosphamide (monohydrate) • Aminoanthracene	• Sodium azide • 2-Nitrofluorene • 9-Aminoacridine • Cumene hydroperoxide • Mitomycin C • N-Ethyl-N-nitro-N-nitrosoguanidine or 4-nitroquinoline 1-oxide • WP2, WP2 uvrA, and WP2 uvrA (pKM101) • Furylfuramide (AF-2)
In vitro mammalian chromosome aberration test	• Methyl methanesulphonate • Ethyl methanesulphonate • Ethylnitrosourea • Mitomycin C • 4-Nitroquinoline-N-oxide	• Benzo(a)pyrene • Cyclophosphamide (monohydrate)
In vitro mammalian cell gene mutation tests using the *Hprt* and *xprt* genes	• *Hprt* 3-Methylcholanthrene • 7,12-Dimethylbenzanthracene • Benzo[a]pyrene • *xprt* benzo[a]pyrene	• *Hprt* Ethylmethanesulfonate • Ethylnitrosourea • 4-Nitroquinoline 1-oxide *xprt* • Streptonigrin • Mitomycin C
In vitro mammalian cell micronucleus test	• Benzo(a)pyrene • Cyclophosphamide • Colchicine • Vinblastine	• Methyl • methanesulphonate • Mitomycin • 4-Nitroquinoline-oxide • Cytosine arabinoside
In vitro mammalian cell gene Mutation tests using the thymidine kinase gene	• Benzo(a)pyrene • Cyclophosphamide (monohydrate) • 7,12-Dimethyl benzanthracene • 3-Methylcholanthrene	• Methyl methanesulphonate • Mitomycin C • 4-Nitroquinoline-N-oxide

DIFFERENT IN VITRO TESTS

Bacterial Reverse Mutation Test (TG 471)

The test derives its name from the fact that these mutations allow the recovery of the normal function of the test strain compared to the parent strain and are hence called "back" or "reverse" mutations; the colonies thus formed are called "revertants" (Fig. 4.2). The bacterial reverse mutation test, or Ames test, uses strains of *Salmonella typhimurium* and *Escherichia coli* that are deficient in an essential amino acid. The mutation is usually present in the gene for the histidine operon in *S. typhimurium*, and in the tryptophan operon for *E. coli* [8]. It is helpful in detecting point mutations such as substitution, addition, or deletion in one or several base pairs of DNA [6,7,9]. These point mutations often develop into various genetic disorders. Upon introduction of a mutagen into the parent strain, reversion of mutation occurs in the test strain and restores the ability of the strain to synthesize the essential amino acid, which its parent strain was not capable of. These revertant bacteria are detected by their ability to grow in the absence of the amino acid, which was crucial for the parent test strain. Bacterial reverse mutation is inexpensive, free of difficulties, and easy to perform.

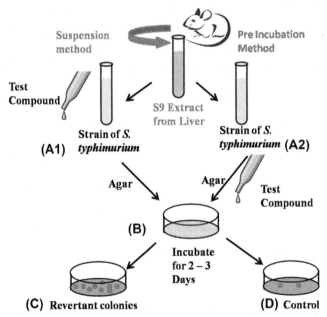

FIGURE 4.2 Bacterial reverse mutation test (Test Guideline [TG] 471). The test is performed either by suspension method (A1) or by preincubation method (A2). The mixture is then spread evenly on the agar plate (B) and allowed to incubate for 2–3 days after which the numbers of revertant colonies (C) versus control colonies (D) are counted.

Principle

A compound likely to be mutagenic in bacteria can be a potent carcinogen in humans. Although only three-fourth of compounds tested are carcinogens in rodents, and certain compounds are rendered negative by this test, its efficacy and low cost make this test an important tool for screening substances for potential carcinogenicity.

Various *S. typhimurium* strains such asTA97, TA98, TA100, TA102, TA104, TA1535, TA1537, and TA1538 are used for testing, as each strain is genetically different. For *E. coli*, the strains used include WP2 uvrA or uvrA(pKM101). The *S. typhimurium* strains must have a GC base at the reversion site and should not be able to detect mutagens such as oxidizing compounds or intercalating agents, whereas the *E. coli* strain should have an AT base pair at the primary reversion site so that the above-mentioned mutagens are recognized by these strains.

Several methods are used in bacterial reverse mutation tests, such as plate incorporation method [6,7,9,10], the preincubation method [11–13], fluctuation method [14,15], and suspension method [16], of which the plate incorporation and preincubation methods are the most common.

A cell suspension is initially exposed to the test substance. In the plate incubation method, this bacterial cell suspension is mixed with a layer of agar, followed by immediate plating onto the minimal medium; in the preincubation method, as the name suggests, the bacterial cell suspension is incubated with the test compound and then mixed with agar before plating to minimal medium. After incubation for 2–3 days, the revertant colonies are counted and compared to the control plate. Short-chain aliphatic compounds like divalent metals, aldehydes, azo dyes, etc., are better detected with the preincubation method [9].

Procedure

The detailed procedure for this test is described in the OECD guidelines [17]. A brief outline follows:

- Fresh bacterial culture (with approximately 10^9 cells, at 37°C) is grown to late exponential or early stationary phase.
- Once the strain of interest is chosen, the bacteria are exposed to the test substance either alone or in the presence of an appropriate metabolic activator as described above.
- Plating is carried out according to the chosen method (see above), the plates are incubated (48–72h, 37°C) and the number of revertant colonies are counted and compared to the control.

Precautions

- Bacterial cells in late stationary phase should be avoided.
- There should be a high titer of viable bacteria.
- For authentication of results, triplicate plating should be used at each test dose.

In Vitro Mammalian Chromosome Aberration Test (TG 473)

The test is used to validate whether a compound is able to introduce structural chromosomal aberrations in cultured mammalian cells [18,19]. The aberrations could be either chromatid-type aberration which includes a single chromatid or chromosome type aberration including both chromatids; the majority of mutagens being chromatid-type aberrations (Fig. 4.3).

The mammalian cells are subjected to a potential mutagen or carcinogen, with or without metabolic activation, and chromosomes are then viewed at metaphase by prior treatment with colcemid [20]. A variety of cell lines, including primary cultures, can be employed for this test.

Procedure

- In brief, an appropriate cell line is chosen and the cells are grown in the required media at optimum temperature, moisture, and CO_2 concentration. The proliferating cells are treated with the test substance, alone, or with a metabolic activator.

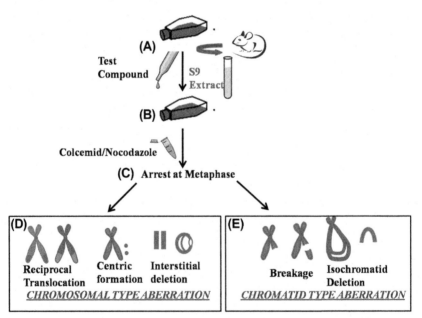

FIGURE 4.3 In vitro mammalian chromosome aberration test (Test Guideline [TG] 473). The cells are plated and treated with the metabolic activator (A), along with the test compound (B). This is followed by exposure of cells to colcemid/nocodazole (C), which allows arrest at metaphase. Various chromosomal type aberration (D) and chromatid type aberration (E) are observed.

- The cells are then exposed for 3–6 h initially and are sampled at 1.5 times normal cell cycle length, after the beginning of treatment [21]. If necessary, cells are treated continuously until sampling at a time equivalent to 1.5 times normal cell cycle length, but without metabolic activation in this case.
- The cells are arrested at metaphase with colcemid, usually 1–3 h before harvest, followed by fixation and staining.

Precautions

- Duplicate cultures should be used with minimal variation in culture.
- Gaseous and volatile compounds should be used in sealed culture vessels to prevent loss of compound [22,23].

In Vitro Mammalian Cell Gene Mutation Tests Using the *Hprt* and *xprt* Genes (TG 476)

This test detects mutation at the locus of hypoxanthine–guanine phosphoribosyl transferase (HPRT) and xanthine–guanine phosphoribosyl transferase transgene (gpt) (XPRT). Usually Chinese hamster lung (V79) or Chinese hamster ovary (CHO) cells are used, which have a stable karyotype of 2N=20 and 2N=21, respectively (Fig. 4.4).

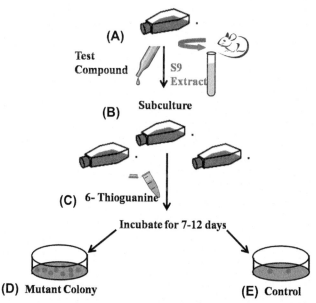

FIGURE 4.4 In vitro mammalian cell gene mutation tests using the Hprt and xprt genes (Test Guideline [TG] 476). The cells are plated and treated with the metabolic activator (A), along with the test compound. The cells are then subcultured to allow the expression of phenotypic characters (B). This is followed by exposure of cells to 6-thioguanine (C) followed by incubation for 7–12 days. The mutant colonies (D) versus the control colonies (E) are then compared.

Principle

There is a mutation in the X chromosome of the gene at the HPRT locus. This locus is normally essential for encoding the HPRT enzyme, a transferase that aids the conversion of hypoxanthine and guanine to inosine monophosphate and guanosine monophosphate, respectively. This enzyme is crucial in the salvage pathway of purine synthesis, and mutation in this gene result in ailments such as the Lesch–Nyhan syndrome [24]. Although both test are equally important, the test using the *xprt* gene is not routinely performed.

Compounds like 6-thioguanine (6-TG) and 8-azaguanine are analogs of the guanine base pair. 6-TG is converted to 6-thioguanosine monophosphate (TGMP), aided by HPRTase. In HPRT+cells, this TGMP (a ribophosphorylated derivative), is cytotoxic for such cells and causes cell death upon incorporation with DNA. High TGMP concentrations accumulate in the cell, rendering the synthesis of guanine nucleotides futile [25]. The cells with functional HPRT or XPRT thus fail to thrive in the presence of 6-TG, as compared to those that lack the locus. When HPRT+cells (like CHO or V79) are incubated with the test compound and replated on 6-TG-containing plates, colony numbers dwindle compared to mutant lines (HPRT−), which, when grown on similar plates, inhibits formation of the toxic metabolite and colony numbers increase.

Procedure

- CHO, CHL, and V79 cell lines are commonly used [26,27] for HPRT tests and CHO-derived AS52 cells are used in the XPRT test. Cells are normally grown in suspension or monolayer, exposed to the test chemical for a stipulated period of time (for example, 3–6 h), and subcultured to check toxic effects of the test chemical as well as phenotypic expression [28–30].
- There must be at least 10 spontaneous mutant frequencies in each culture at each test stage according to standard protocol [31], and the spontaneous frequency is generally around 5×10^{-6}.
- The relative survival (RS) is the measure of cytotoxicity. The treated cultures are allowed to grow in the medium for 7–9 days, thereby allowing sufficient time for the expression of any phenotype induced by the test compound.
- Once phenotypic expression appears, the cells are replated in the medium alone or with the selective agent 6-thioguanine to test the cloning efficiency and the number of mutant strains.
- Plates are incubated for a period of 7–12 days to allow the cells to proliferate in the selective conditions, and colonies are then counted. Mutant frequency is then calculated from the formula:

$$\text{Mutant frequency} = \frac{\text{Cloning efficiency of mutant colonies in selective medium}}{\text{Cloning efficiency in non-selective medium}}$$

Precautions

- The cell line selected should be highly sensitive to chemical mutagens, have a stable karyotype and high cloning efficiency.
- Cell lines should be tested frequently for the absence of mycoplasma contamination and for modal chromosome number.

In Vitro Mammalian Cell Micronucleus Test (TG 487)

This assay is used to detect the presence of micronuclei in interphase cells. The term micronucleus refers to a "small nucleus" (Fig. 4.5).

Principle

Whenever a cell divides, it produces daughter cells with equal genetic content due to the simultaneous division of the genetic material. If the process is disturbed by endogenous or exogenous factors, however, the chromosomes are damaged, often broken, and this affects the distribution of the genetic content in daughter cells.

In such a scenario, the damaged portion is not incorporated in the newly formed nucleus and is retained in the cell as a micronucleus. A simple way to find whether a compound is genotoxic or not is to treat the cells with that

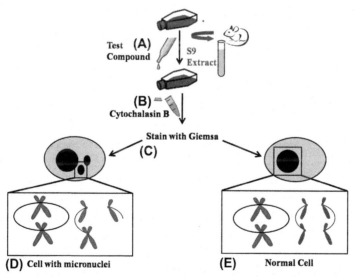

FIGURE 4.5 In vitro mammalian cell micronucleus test (Test Guideline [TG] 487). The cells are plated and treated with the metabolic activator (A), along with the test compound. The cells are then treated with cytochalasin B (B) to prevent cytokinesis. The cells are then stained with Giemsa stain (C) and the micronuclei are visualized (D) under microscope and compared to the normal cells (E).

compound and check the micronucleus formation rate. Micronuclei might develop from acentric or centric chromosomes that failed to migrate during the cell anaphase [32,33]. Micronucleus formation is, therefore, the indication of damage incurred on the daughter cells.

This is a robust and excellent test for testing the genotoxicity of compounds and can be performed in a number of cell lines [34–36].

Procedure

- The cell line of interest is chosen and treated with the test chemicals for 3–6h. This is followed by removal of the test compound. The cells are then treated with cytochalasin B to block cytokinesis, followed by sampling at a time equivalent to about 1.5–2.0 normal cell cycle duration [37].
- The different sets of cell cultures are harvested and processed separately.
- Staining of cells is carried out with various dyes such as Giemsa or fluorescent DNA-specific dyes, followed by fixation.
- Cells are then viewed under a microscope and micronuclei are checked.

Precaution

Cells should be plated in triplicate to eliminate any possibility of false-negative result.

In Vitro Mammalian Cell Gene Mutation Tests Using the TK Gene (TG 490)

This test is used to detect mutation caused by a chemical. Reporter genes such as TK are used to test the forward mutation. Commonly used cell lines are L5178Y TK±3.7.2 C mouse lymphoma (generally called L5178Y) and the TK6 human lymphoblastoid (generally termed TK6). These cells might differ in their characteristic cell growth, origin, and other factors; nonetheless, the assay protocol is similar for both lines.

Cells with a deficiency in the TK enzyme (due to a mutation from TK± to TK−/−) often arise due to chromosomal aberration or gene mutation. Gene mutations include point mutations, frameshift mutation, and deletions, whereas chromosomal aberrations can be due to lack of a large portion of chromosomal DNA or rearrangements, either of which leads to abnormal chromosome numbers. If there is nondisjunction during the mitotic event or spindle impairment, an entire chromosome could be lost. The common drawback of this test is that it fails to recognize the aneugens (compounds that render faulty or abnormal numbers of chromosomes in daughter chromosomes) and thus cannot be used to detect them (Fig. 4.6) [38–40].

When the assay is performed, two phenotypes are often formed; the normal-growing mutants, which grow at the same rate as the TK heterozygous cells, and the slow-growing mutants, which have an extended doubling time compared

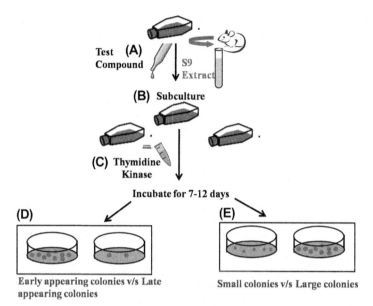

FIGURE 4.6 **In vitro mammalian cell gene mutation tests using the thymidine kinase (TK) gene (Test Guideline [TG] 490).** The cells are plated and treated with the metabolic activator (A), along with the test compound. The cells are then subcultured to allow the expression of phenotypic characters (B). This is followed by exposure of cells to TK (C) followed by incubation for 7–12 days. The number of early appearing versus late appearing mutants colonies are compared (D) along with the small and the large colonies (E).

to TK heterozygous cells. For mouse lymphomas, they are termed large and small colonies, whereas for the TK6 assay they are regarded as early and late-appearing colonies. The details of these are procedures are dealt with separately [41–45].

When there is damage near the TK locus, it results in prolonged doubling time and thus, formation of a small colony, which obviously would appear late. Similarly, if the damage is not near the TK locus, cell doubling time is normal and they appear as colonies at a time similar to that of the parental colony. It can thus be inferred that in the former case, damage is due to structural changes at the chromosome level, whereas in the latter case, damage is due to point mutations [41,43,46].

Principle

Cells are grown in suspension or as a monolayer and exposed to the test chemical, alone or with a metabolic activator; they are then subcultured for the prerequisite time such that the phenotypic expression appears. Cytotoxicity is evaluated by relative growth of the mouse lymphoma assay (MLA) and RS for TK6. The treated cells are then grown in the media without the insult and allowed to grow until the phenotypic expression is apparent, which is followed by determination of mutant frequency by seeding a known number of cells in the medium with a selective agent (to determine the mutant colonies) and without it (to confirm the viable colonies) and incubated, after which the colonies are counted.

MLA uses TK+/− −3.7.2C subline of L5178Y cells. The line is derived from a methylcholanthrene-induced thymic lymphoma from a DBA-2 mouse [47]. The modal chromosomal number is with one metacentric chromosome (t12; 13) and has a published karyotype [48–51]. These cells have both p53 mutant alleles.

TK6 are human lymphoblastoid cells, derived from an Epstein–Barr virus-transformed cell line, WI-L2. The cells are diploid with karyotype 47; XY, 13+, t(14; 20), t(3; 21) [52]. TK6 is a p53-competent cell line [53].

Procedure

- Proliferating cells are treated with the test chemical alone or in the presence of a metabolic activator, for a period of 3–4 h. If the cell (specifically MLA cells) response is negative, the duration is increased to 24 h without S9.
- Once this treatment period is over, cells are cultured for a given time to allow expression of the phenotype (for MLA the duration is 2 days, whereas for TK6 it is 3–4 days); cells are counted daily.
- Cells are then suspended in the medium with or without a selective agent (triflurothymidine for TK mutants [42]) and the number of mutants is determined as well as the cloning efficiency, respectively.
- The cells are then plated on soft agar or in 96-well plates (for MLA) or in liquid media and 96-well plates (for TK6).
- Once plated, the cells are counted after 10–14 days, depending on cell type, and necessary calculations are made to determine the mutation frequency.
- Colony size or time taken for colony appearance is counted.

CURRENT LIMITATIONS OF IN VITRO GENOTOXICITY TESTING

Although there are many in vitro tests available to assess whether or not a given compound is mutagenic, these methods often give false positive results. Certain major drawbacks of in vitro tests include the following:

Faulty Metabolic Activator System

- The test substance often must be metabolically activated. The level of metabolic efficiency is directly related to the degree of organization, i.e., whole organ, organ slices, enzymatically active primary cell culture, and so on. The commonly used activator is the S9 system, which might influence the result, depending upon various parameters.

- "Inducer effect": metabolic induction strongly affects the expression of CYP 450, which itself can be activated by various compounds such as phenobarbital, 5.6-benzoflavone, the polychlorinated biphenyl mixture Aroclor 1254, etc. [54]. A single compound might lower one form of cytochrome p450, while elevating levels of others. There is also variation in normal CYP450 levels in human liver compared to that in rat liver.
- The age, sex [55,56] species (rodent, dog, or human) [57,58], as well as the organ origin [58] play a vital role in the metabolizing effect and thus the mutagenic level of the drug of interest.
- To overcome these issues, genetically engineered cells were used that express the human or rat p450 only, but they turned out to be more unreliable and gave false results [59].Certain compounds, when activated by human liver S9, were less mutagenic than the untreated rat livers, whereas the mutagenic potency of the same drugs increased when treated with Aroclor.
- There is thus no permanent, full proof system for metabolic activation and every system has limitations; the test must thus be verified using in vivo methods.

False Positive Results

- In the Ames test, *S. typhimurium* is used, which has certain enzymes such as like nitroreductases that are absent in humans. The nitro compounds are thus often regarded as mutagens in the bacterial system, whereas in reality they might not be mutagens in humans. Also in the Ames test, mutations permit the test colonies to grow in the absence of histidine, but certain test compounds can contain amino acids such as arginine or even histidine itself, which can promote growth of nonmutant colonies [60,61].
- Mammalian cell lines can also give faulty results. The CHO-K1 and CHO WBL cell lines used in in vitro tests have a faulty p53 sequence and lack the G1 checkpoint. Similarly, CHL cells have an intact p53 sequence but expression is often faulty [62].
- Certain compounds that involve micronucleus formation in the cells did not produce similar results when tested in in vivo systems [63].

False Negative Results

The Ames test is rendered negative when nanomaterials are used as bacterial cells probably form a barrier to nanomaterials [64].

There are thus various limitations with the in vitro test system, and almost every assay has some drawback and limitations. Before concluding whether a compound is genotoxic or not, extensive tests should be done to validate the results in the in vivo system. There should be thorough investigation to determine whether a compound that renders a positive or negative results in the test system (in vitro) that is actually the same, or is merely a consequence of faulty interaction with the test system itself.

There is another in vitro test called the Comet assay or single-cell gel electrophoresis assay, which was originally developed by Ostling and Johanson [65] and later was optimized by Singh et al. [66] and Olive and Banath [67]. This assay can detect various forms of DNA damage and is rapid, sensitive, and relatively simple to detect. Despite its promising genotoxicity potential, there is no official international guideline available for its conduct. However, the International Workshop on Genotoxicity Test Procedure in 1999 [68] has proposed standard operating protocols for the same, nonetheless till date OECD has developed only in vivo comet assay protocol (TG 489). Therefore, in vitro comet assay test is widely used in basic research only. In pharmaceutical and chemical industry, the test is used as a screening assay rather than for regulatory purpose in market authorization or clinical trials.

FUTURE PROSPECTIVE OF THE GENOTOXICITY TESTING

Though at present most of the toxicological studies are based on traditional animal experiments having acceptance of regulatory authorities, still they are not always the best choice for predicting human health effects [69–71]. A tiered battery test of genotoxicity test would help in the selection of drugs and chemicals for human use. Over the last decades, cell culture methodologies have substantially improved resulting in a number of promising cell models composed of relevant morphological and biochemical signaling processes that may reproduce the in vivo environment [72]. The genotoxicity tests are an important component required for the safety assessment of chemicals and drugs in compliance with regulatory guidelines set by international regulatory authorities such as European Union to protect the human health and environment.

Despite all the above advancements, the genotoxic methods fail to completely replace the current traditional tests across all sectors. The novel in vitro model system along with other emerging technologies such as high-throughput assay system, imaging technologies, computational approaches will help in developing completely new integrated approaches for genotoxicity testing and assessment, which will be more advanced and promising than the traditional methods established over the years [73]. In the current scenario, more research, development, and data integration are required to meet the demands of regulatory testing. Only then, there is a possible likelihood for reduction and complete replacement of traditional testing method acceptable to regulatory authorities.

SUMMARY

Everyday, new drugs are invented whose basic toxicological profiles have to be checked. Therefore, genotoxicity testing becomes pivot and prerequisite for any new drug and is a mandatory protocol by the regulatory authorities. OECD has set up various guidelines for different genotoxicity tests, which are very much required so as to get acceptance of a new drug in the market. A standard test battery of genotoxicity test composed of two in vitro tests (one for bacterial and other for mammalian cells) and one in vivo test is required to validate the efficacy of a drug. Though the genotoxicity methods are simple, less time consuming and sensitive than the traditional animal tests in all manners, they fail to completely replace the former ones due to certain limitations. So it is assumed that the present genotoxic methods along with other emerging technologies will help in developing completely new integrated approaches for genotoxicity testing and assessment, which will be more advanced and promising than the traditional methods established over the years.

ACKNOWLEDGMENTS

The authors thank Ms. Poorwa Awasthi and Mr. Vipin Yadav for their critical reading of the chapter and editorial assistance. Work in A.K. Lab is supported by the CSIR, Indepth and, Epigenetics in health and disease network projects and funding from Department of Science and Technology, Govt. of India. The work in A.K.'s lab is also funded by Wellcome Trust and DBT India Alliance. IITR manuscript transmission number is 3458.

REFERENCES

[1] Wogan GN, Hecht SS, Felton JS, Conney AH, Loeb LA. Environmental and chemical carcinogenesis. Seminars Cancer Biol December 2004;14(6):473–86.

[2] Phillips DH, Arlt VM. Genotoxicity: damage to DNA and its consequences. Exs Suppl 2009;99:87–110.

[3] De Flora S, Izzotti A. Mutagenesis and cardiovascular diseases Molecular mechanisms, risk factors, and protective factors. Mutat Res August 01, 2007;621(1–2):5–17.

[4] Kirkland D, Kasper P, Martus HJ, Muller L, van Benthem J, Madia F, et al. Updated recommended lists of genotoxic and non-genotoxic chemicals for assessment of the performance of new or improved genotoxicity tests. Mutat Res Genet Toxicol Environ Mutagen January 01, 2016;795:7–30.

[5] OECD. Version OECD guideline for the testing of chemicals new test guideline; in vitro mammalian cell gene mutation assays using the thymidine kinase gene. 2014.

[5a] ICH. ICH guideline S2 (R1) on genotoxicity testing and data interpretation for pharmaceuticals intended for human use; 2012.

[6] Ames BN, McCann J, Yamasaki E. Methods for detecting carcinogens and mutagens with the Salmonella/mammalian-microsome mutagenicity test. Mutat Res December 1975;31(6):347–64.

[7] Maron DM, Ames BN. Revised methods for the Salmonella mutagenicity test. Mutat Research May 1983;113(3–4):173–215.

[8] Elliott BM, Combes RD, Elcombe CR, Gatehouse DG, Gibson GG, Mackay JM, et al. Alternatives to Aroclor 1254-induced S9 in in vitro genotoxicity assays. Mutagenesis May 1992;7(3):175–7.

[9] Gatehouse D, Haworth S, Cebula T, Gocke E, Kier L, Matsushima T, et al. Recommendations for the performance of bacterial mutation assays. Mutat Res June 1994;312(3):217–33.

[10] Kier LD, Brusick DJ, Auletta AE, Von Halle ES, Brown MM, Simmon VF, et al. The *Salmonella typhimurium*/mammalian microsomal assay. A report of the U.S. Environmental Protection Agency Gene-Tox Program. Mutat Research September 1986;168(2):69–240.

[11] Yahagi T, Degawa M, Seino Y, Matsushima T, Nagao M. Mutagenicity of carcinogenic azo dyes and their derivatives. Cancer Lett November 1975;1(2):91–6.

[12] Matsushima M, Sugimura T, Nagao M, Yahagi T, Shirai A, Sawamura M. Factors modulating mutagenicity microbial tests. In: Short-term test systems for detecting carcinogens. Springer; 1980. p. 273–85.

[13] Aeschbacher HU, Wolleb U, Porchet L. Liquid preincubation mutagenicity test for foods. J Food Saf 1987;8:167–77.

[14] Green MH, Muriel WJ, Bridges BA. Use of a simplified fluctuation test to detect low levels of mutagens. Mutat Res February 1976;38(1):33–42.

[15] Hubbard SA, Green MHL, Gatehouse D, Bridges JW. The fluctuation test in bacteria. Handbook of mutagenicity test procedures (2)1984.

[16] Thompson ED, Melampy PJ. An examination of the quantitative suspension assay for mutagenesis with strains of *Salmonella typhimurium*. Environ Mutagenesis 1981;3(4):453–65.

[17] OECD G. OECD guideline for testing of chemicals. 2016. p. 471.

[18] Evans HJ. Cytological methods for detecting chemical mutagens. In: Chemical mutagens, principles and methods for their detection;4. 1976. p. 1–29.

[19] Galloway SM, Armstrong MJ, Reuben C, Colman S, Brown B, Cannon C, Bloom AD, Nakamura F, Ahmed M, Duk S, Rimpo J, Margolin GH, Resnick MA. Chromosome aberration and sister chromatid exchanges in Chinese hamster ovary cells: evaluation of 108 chemicals. Environ Molec Mutagen 1987;10:1–175.

[20] OECD. OECD guideline for the testing of chemicals in vitro mammalian chromosome aberration test. 2016.

[21] Galloway SM, Aardema MJ, Ishidate Jr M, Ivett JL, Kirkland DJ, Morita T, et al. Report from working group on in vitro tests for chromosomal aberrations. Mutat Res June 1994;312(3):241–61.

[22] Krahn DF, Barsky FC, McCooey KT. CHO/HGPRT mutation assay: evaluation of gases and volatile liquids. 1982. p. 91–103.

[23] Zamora PO, Benson JM, Li AP, Brooks AL. Evaluation of an exposure system using cells grown on collagen gels for detecting highly volatile mutagens in the CHO/HGPRT mutation assay. Environ Mutagenesis 1983;5(6):795–801.

[24] Nguyen KV, Nyhan WL. Identification of novel mutations in the human HPRT gene. Nucleosides Nucleotides Nucleic Acids 2013;32(3):155–60.

[25] Evans WE. Pharmacogenetics of thiopurine S-methyltransferase and thiopurine therapy. Ther Drug Monit April 2004;26(2):186–91.

[26] Tindall KR, Stankowski Jr LF, Machanoff R, Hsie AW. Detection of deletion mutations in pSV2gpt-transformed cells. Mol Cell Biol July 1984;4(7):1411–5.

[27] Hsie AW, Recio L, Katz DS, Lee CQ, Wagner M, Schenley RL. Evidence for reactive oxygen species inducing mutations in mammalian cells. Proc Natl Acad Sci USA December 1986;83(24):9616–20.

[28] Li AP, Carver JH, Choy WN, Hsie AW, Gupta RS, Loveday KS, et al. A guide for the performance of the Chinese hamster ovary cell/hypoxanthine-guanine phosphoribosyl transferase gene mutation assay. Mutat Res October 1987;189(2):135–41.

[29] Liber HL, Yandell DW, Little JB. A comparison of mutation induction at the tk and hprt loci in human lymphoblastoid cells; quantitative differences are due to an additional class of mutations at the autosomal tk locus. Mutat Res February 1989;216(1):9–17.

[30] Stankowski Jr LF, Tindall KR, Hsie AW. Quantitative and molecular analyses of ethyl methanesulfonate- and ICR 191-induced mutation in AS52 cells. Mutat Res April 1986;160(2):133–47.

[31] O'Neill JP, Hsie AW. Phenotypic expression time of mutagen-induced 6-thioguanine resistance in Chinese hamster ovary cells (CHO/HGPRT system). Mutat Res January 1979;59(1):109–18.

[32] OECD. Overview of the set of OECD genetic toxicology test guidelines and updates performed in 2014–2015. 2016.

[33] Kirsch-Volders M. Towards a validation of the micronucleus test. Mutat Res 1997;392:1–4.

[34] Clare MG, Lorenzon G, Akhurst LC, Marzin D, van Delft J, Montero R, et al. SFTG international collaborative study on in vitro micronucleus test II. Using human lymphocytes. Mutat Res August 04, 2006;607(1):37–60.

[35] Aardema MJ, Snyder RD, Spicer C, Divi K, Morita T, Mauthe RJ, et al. SFTG international collaborative study on in vitro micronucleus test III. Using CHO cells. Mutat Res August 04, 2006;607(1):61–87.

[36] Wakata A, Matsuoka A, Yamakage K, Yoshida J, Kubo K, Kobayashi K, et al. SFTG international collaborative study on in vitro micronucleus test IV. Using CHL cells. Mutat Res August 04, 2006;607(1):88–124.

[37] Fenech M, Morley AA. Cytokinesis-block micronucleus method in human lymphocytes: effect of in vivo ageing and low dose X-irradiation. Mutat Res July 1986;161(2):193–8.

[38] Fenech MD, Luker T, Cooper A, O'Donovan MR. Unusual structure- genotoxicity relationship in mouse lymphoma cells observed with a series of kinase inhibitors. Mutat Res 2012;746(1):21–8.

[39] Honma M, Momose M, Sakamoto H, Sofuni T, Hayashi M. Spindle poisons induce allelic loss in mouse lymphoma cells through mitotic non-disjunction. Mutat Res June 27, 2001;493(1–2):101–14.

[40] Wang J, Sawyer JR, Chen L, Chen T, Honma M, Mei N, et al. The mouse lymphoma assay detects recombination, deletion, and aneuploidy. Toxicol Sci May 2009;109(1):96–105.

[41] Applegate ML, Moore MM, Broder CB, Burrell A, Juhn G, Kasweck KL, et al. Molecular dissection of mutations at the heterozygous thymidine kinase locus in mouse lymphoma cells. Proc Natl Acad Sci USA January 1990;87(1):51–5.

[42] Moore MM, Clive D, Hozier JC, Howard BE, Batson AG, Turner NT, et al. Analysis of trifluorothymidine-resistant (TFTr) mutants of L5178Y/TK+/- mouse lymphoma cells. Mutat Res August 1985;151(1):161–74.

[43] Hozier J, Sawyer J, Clive D, Moore MM. Chromosome 11 aberrations in small colony L5178Y TK-/- mutants early in their clonal history. Mutat Res October 1985;147(5):237–42.

[44] Honma M, Hayashi M, Sofuni T. Cytotoxic and mutagenic responses to X-rays and chemical mutagens in normal and p53-mutated human lymphoblastoid cells. Mutat Res March 04, 1997;374(1):89–98.

[45] Honma M, Momose M, Tanabe H, Sakamoto H, Yu Y, Little JB, et al. Requirement of wild-type p53 protein for maintenance of chromosomal integrity. Mol Carcinog August 2000;28(4):203–14.

[46] Schisler MR, Moore MM, Gollapudi BB. In vitro mouse lymphoma (L5178Y Tk(+)/(-)-3.7.2C) forward mutation assay. Methods Mol Biol 2013;1044:27–50.

[47] Fischer GA. Studies of the culture of leukemic cells in vitro. Ann NY Acad Sci December 05, 1958;76(3):673–80.

[48] Sawyer J, Moore MM, Clive D, Hozier J. Cytogenetic characterization of the L5178Y TK+/-3.7.2C mouse lymphoma cell line. Mutat Res October 1985;147(5):243–53.

[49] Sawyer JR, Moore MM, Hozier JC. High-resolution cytogenetic characterization of the L5178Y TK+/- mouse lymphoma cell line. Mutat Res October 1989;214(2):181–93.

[50] Sawyer JR, Binz RL, Wang J, Moore MM. Multicolor spectral karyotyping of the L5178Y Tk+/- -3.7.2C mouse lymphoma cell line. Environ Molecular Mutagenesis March 2006;47(2):127–31.

[51] Fellows MD, McDermott A, Clare KR, Doherty A, Aardema MJ. The spectral karyotype of L5178Y TK(+)/(-) mouse lymphoma cells clone 3.7.2C and factors affecting mutant frequency at the thymidine kinase (tk) locus in the microtitre mouse lymphoma assay. Environ Mol Mutagen January 2014;55(1):35–42.

[52] Honma M. Generation of loss of heterozygosity and its dependency on p53 status in human lymphoblastoid cells. Environ Mol Mutagen March–April 2005;45(2–3):162–76.

[53] Xia F, Wang X, Wang YH, Tsang NM, Yandell DW, Kelsey KT, et al. Altered p53 status correlates with differences in sensitivity to radiation-induced mutation and apoptosis in two closely related human lymphoblast lines. Cancer Res January 01, 1995;55(1):12–5.

[54] Guengerich FP, Dannan GA, Wright ST, Martin MV, Kaminsky LS. Purification and characterization of liver microsomal cytochromes p-450: electrophoretic, spectral, catalytic, and immunochemical properties and inducibility of eight isozymes isolated from rats treated with phenobarbital or beta-naphthoflavone. Biochemistry November 09, 1982;21(23):6019–30.

[55] Imaoka S, Fujita S, Funae Y. Age-dependent expression of cytochrome P-450s in rat liver. Biochim Biophys Acta October 21, 1991;1097(3):187–92.

[56] Kamataki T, Maeda K, Yamazoe Y, Nagai T, Kato R. Sex difference of cytochrome P-450 in the rat: purification, characterization, and quantitation of constitutive forms of cytochrome P-450 from liver microsomes of male and female rats. Arch Biochem Biophys September 1983;225(2):758–70.

[57] Cox JA, Fellows MD, Hashizume T, White PA. The utility of metabolic activation mixtures containing human hepatic post-mitochondrial supernatant (S9) for in vitro genetic toxicity assessment. Mutagenesis March 2016;31(2):117–30.

[58] Nishimuta H, Nakagawa T, Nomura N, Yabuki M. Species differences in hepatic and intestinal metabolic activities for 43 human cytochrome P450 substrates between humans and rats or dogs. Xenobiotica November 2013;43(11):948–55.

[59] Nesslany F. The current limitations of in vitro genotoxicity testing and their relevance to the in vivo situation. Food Chem Toxicol August 31, 2016. http://dx.doi.org/10.1016/j.fct.2016.08.035.

[60] Khandoudi N, Porte P, Chtourou S, Nesslany F, Marzin D, Le Curieux F. The presence of arginine may be a source of false positive results in the Ames test. Mutat Res September–Octber 2009;679(1–2):65–71.

[61] Couderc T, Khandoudi N, Grandadam M, Visse C, Gangneux N, Bagot S, et al. Prophylaxis and therapy for Chikungunya virus infection. J Infect Dis August 15, 2009;200(4):516–23.

[62] Chaung W, Mi LJ, Boorstein RJ. The p53 status of Chinese hamster V79 cells frequently used for studies on DNA damage and DNA repair. Nucleic Acids Res March 01, 1997;25(5):992–4.

[63] Honma M, Hayashi M. Comparison of in vitro micronucleus and gene mutation assay results for p53-competent versus p53-deficient human lymphoblastoid cells. Environ Mol Mutagen June 2013;52(5):373–84.

[64] Nesslany F, Simar-Meintieres S, Watzinger M, Talahari I, Marzin D. Characterization of the genotoxicity of nitrilotriacetic acid. Environ Mol Mutagen July 2008;49(6):439–52.

[65] Ostling O, Johanson KJ. Microelectrophoretic study of radiation-induced DNA damages in individual mammalian cells. Biochem Biophys Res Commun August 30, 1984;123(1):291–8.

[66] Singh NP, McCoy MT, Tice RR, Schneider EL. A simple technique for quantitation of low levels of DNA damage in individual cells. Exp Cell Res March 1988;175(1):184–91.

[67] Olive PL, Banath JP. The comet assay: a method to measure DNA damage in individual cells. Nat Protoc 2006;1(1):23–9.

[68] Tice RR, Agurell E, Anderson D, Burlinson B, Hartmann A, Kobayashi H, et al. Single cell gel/comet assay: guidelines for in vitro and in vivo genetic toxicology testing. Environ Mol Mutagen 2000;35(3):206–21.

[69] Olson H, Betton G, Robinson D, Thomas K, Monro A, Kolaja G, et al. Concordance of the toxicity of pharmaceuticals in humans and in animals. Regul Toxicol Pharmac August 2000;32(1):56–67.

[70] Bracken MB. Why animal studies are often poor predictors of human reactions to exposure. J R Soc Med March 2009;102(3):120–2.

[71] Hartung T, Koeter H. Food for thought... on food safety testing. Altex 2008;25(4):259–64.

[72] Carere A, Stammati A, Zucco F. In vitro toxicology methods: impact on regulation from technical and scientific advancements. Toxicol Lett February 28, 2002;127(1–3):153–60.

[73] Zhang L, McHale CM, Greene N, Snyder RD, Rich IN, Aardema MJ, et al. Emerging approaches in predictive toxicology. Environ Mol Mutagen December 2010;55(9):679–88.

Chapter 5

In Silico Approaches for Predictive Toxicology

Ramakrishnan Parthasarathi, Alok Dhawan
CSIR-Indian Institute of Toxicology Research, Lucknow, India

INTRODUCTION

Chemical utilization is on ever-increasing trend in everyday life. For instance, more than 80,000 commercial products with myriad applications are intensely linking one into the world of chemicals [1]. Industrialization and by various means of technological advancements and intervenes, chemicals are posing the major source of threat toward several health and the environmental issues. Understanding and characterization of chemical properties and its interactions with biological systems are essential for the safe use of any existing and emerging chemicals/drugs/materials [2]. Due to the complexity of various toxicological endpoints to ensure the safety of chemicals, a variety of techniques are being performed [3]. Currently available methods to determine the safety of chemicals are mainly concentrating on battery of toxicological testing and assessment using animals [3–5]. However, great efforts have been instigated for developing alternatives to the traditional toxicity testing regime [6]. Hence, paradigm shift in regular practice is needed in terms of proper validation and acceptance of alternatives toxicology evaluation as replacements. In silico toxicology is one of the alternatives to animal testing that complement in vitro and in vivo toxicity assessments to potentially minimize animal testing, reduce the cost and time, and improve toxicity prediction and safety assessment [3,4,7]. In addition, in silico toxicity models have a unique advantage of being able to predict toxicity of chemicals prior to synthesis and to innovate the process of developing new chemical entities with desired properties [2,6].

IN SILICO TOXICOLOGY FRAMEWORK FOR TOXICITY/SAFETY PREDICTION

Toxicity is a measure of potential adverse effects caused by chemicals on humans, animals, plants, or the environment through acute-exposure (single dose)

In Vitro Toxicology. http://dx.doi.org/10.1016/B978-0-12-804667-8.00005-5

or multiple-exposure (multiple doses). The goal of toxicity testing is to develop quantifiable information for enabling adequate precaution and protection for the societal health against adverse effects of various chemicals, and drugs. [8–10]. Adverse effects of chemicals are determined by toxicity endpoints, such as carcinogenicity or genotoxicity (quantitative: lethal dose to 50% (LD_{50}), effective concentration (EC_{50}), lethal concentration (LC_{50}), concentration to inhibit growth (IGC_{50}); qualitative: toxic or nontoxic, low, moderate, and high toxicity) [11]. List of factors such as route of exposure (oral, dermal, inhalation, etc.), dose (concentration), frequency/duration of exposure, ADME properties (absorption, distribution, metabolism, and excretion/elimination), physiology and biological properties (weight, age, gender), and chemical properties influence the toxicity of chemicals [1,12]. In vivo animal toxicity testing (typically rodents, dogs, and/or nonhuman primates) has been considered as the standard for identifying potential adverse effects of chemicals [13]. However, information on the mode and the mechanism of toxicity and differences in toxicological responses of animal data to humans are few concerns on in vivo approaches. In vitro toxicity tests have become more attractive and feasible due to the advances in high-throughput screening, require minimal quantities of testing compounds and reduce animal testing (Reduce, Replace, and Refine animal testing [the 3R principles]) [14,15]. In vitro toxicity assays are conducted primarily with cells or cell lines, ideally from humans or transfected prokaryotic cells for mechanistic investigations. In vitro toxicity methods are well suited for initial screening of toxicity and is to combine predictions with in silico models for overcoming the impact on interpretation of selective endpoints, discovery, and decision space [16–18].

In silico toxicology is a broad term multidisciplinary area; in general computational toxicity assessment, virtual screening, and predictive platform that uses computational resources (methods, algorithms, software, data, etc.) to organize, analyze, model, simulate, visualize, or predict toxicity of chemicals/drugs (Fig. 5.1) [17,19]. It is intertwined with physics, chemistry, biology, mathematics, computer science, and informatics with toxicology, which uses information from computational tools to analyze beneficial or adverse effects envisaging toxicity endpoints of compounds [11]. It is important to highlight here that a variety of computational technique, which relate the structure of a chemical to its toxicity or end effects. In silico toxicology is a predictive technique that helps in retrieving relevant data and/or make predictions regarding the effects of chemicals. There are, obviously, many advantages of computational techniques: in silico predictive toxicology is used in combination with in vitro and in vivo experimental data obtained at the molecular, cellular, organ, organism, and population levels, provide the possibility of improved safety at molecular and functional changes occurring across multiple levels of biotic organization to characterize and evaluate interactions between potential hazards with the components of biological system [20–23]. Predictive toxicology, therefore,

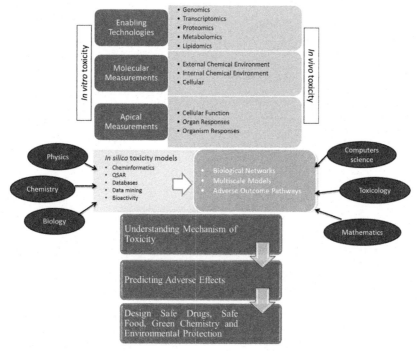

FIGURE 5.1   In silico toxicology framework for toxicity/safety prediction of chemicals/drugs and integrated in vivo and in vitro methodology for a priori risk assessment.

has an ultimate potential for extrapolating quantifiable chemical toxicity endpoints (Table 5.1), and its application could be part of a new paradigm for risk assessment.

In Silico Toxicology

Several in silico approaches have been developed for predictive toxicology. In silico tools vary in complexity and performance and can be broadly classified into four major categories (Fig. 5.2). (1) Structure activity modeling: quantitative structure activity relationships (QSAR), expert systems, grouping and read-across techniques; (2) chemoinformatics: generating molecular descriptors using computational tools including quantum chemical methods and molecular dynamics simulations for toxicity prediction; (3) databases and gathering biological data that contain relations between chemicals and toxicity endpoints, databases for storing data about chemicals, toxicity and chemical properties; (4) data mining and analysis: calculating molecular descriptors, generating a prediction model, evaluating the accuracy and interpreting the model, statistical methods and prebuilt models in web servers or standalone applications for

TABLE 5.1 In Silico Tools Used for Predicting Toxicity Endpoints of Chemicals/Drugs

In Silico Methods	Description	Software/ Databases
Quantitative structure activity relationship models	Use molecular descriptors to predict chemical toxicity	OECD QSAR
		TopKat
		Derek Nexus
		VEGA
		METEOR
		vLife-QSARpro
Structural alerts and rule-based models	Chemical structures that indicate or associate to toxicity	OECD QSAR
		Toxtree
		OCES
		Derek Nexus
		HazardExpert
		Meteor
		CASE
		PASS
		cat-SAR
Read-across	Predicting unknown toxicity of a chemical using similar chemicals with known toxicity from the same chemical category	OECD QSAR
		Toxmatch
		ToxTree
		AMBIT
		AmbitDiscovery
		AIM
		DSSTox
		ChemIDplus
Dose–response and time–response models	Relation between doses (or time) and the incidence of a defined biological effect.	CEBS
		PubChem
		ToxRefDB
Pharmacokinetic (PK) and Pharmacodynamics (PD) models	PK models calculate concentration at a given time. PD models calculate effect at a given concentration	WinNonlin
		Kinetica
		ADAPT

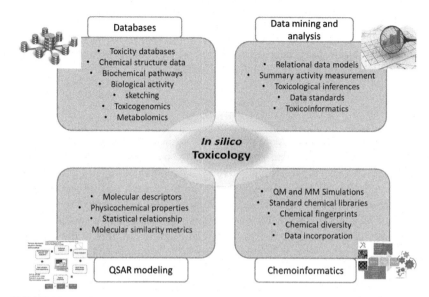

FIGURE 5.2   Multidimensional and broadly integrated in silico categories of predictive toxicology.

predicting toxicity. There are various procedures to unravel general or specific compound toxicity/safety and each method has respective strengths, limitations, the scope of application and interpretation [9]. The underlying principle is to find the suitable and the most effective method to address the particular issue. However, all the four categories mentioned in this section for in silico predictive tools are highly interrelated. The scope of this chapter covers the QSAR, the state-of-the-art method for generating toxicity prediction models in association with descriptors used to form correlation with in vivo/in vitro toxicity endpoints and also providing a highlight on diverse and/or large data sets of toxicology databases and online/standalone computational tools utilized for predictive toxicology.

QSAR FOR PREDICTIVE TOXICOLOGY

The QSAR, quantitative structure property relationship, and quantitative structure toxicity relationship are the important tools of the bio-chemo-informatics, which can be constructed primarily based on the data generated from the molecular modeling and computational chemistry. In general, QSAR framework attempts to find a mathematical relationship between chemical structure and biological activity or chemical property including toxicity for a series of compounds [24]. These series of compounds are called the training set. The generated mathematical equation can be used to predict the activity or property of any new compound, which has been built from the chosen training set. It is noteworthy to mention that the seminal contribution made by Hansch and coworkers for the development of QSAR area of predictive activity [25].

Eq. (5.1) portraits the definition of QSAR, as that a given biological activity can be correlated with the physicochemical properties (size/shape parameters, lipophilicity, electronic properties, structural information) of a compound using a quantitative mathematical relationship [26,27].

$$\log\frac{1}{C} = 4.1\pi - 2.1\pi^2 + 2.8\sigma + 3.4$$

(5.1)

where C is the concentration to produce herbicidal effect, π is an indicator of hydrophobicity, and σ is a measure of electronic effects within the molecule. Many detailed descriptions of QSAR methods and development have been published in the literature [27,28]. Here, the emphasis is on the current state-of-the-art, with comment on the future of QSAR and its potential utility in predictive toxicology.

The fundamental approach for developing, validating, and using QSARs is moderately consistent. QSAR is a technique that tries to predict the activity, reactivity, and properties of an unknown set of molecules based on analysis of an equation connecting the structures of molecules to their respective measured activity and property. Generally, at first the variation in activity and/or property and/or toxicity of a known set of structurally analog samples is studied with the changes in their molecular frameworks. The trends obtained from the study are then transformed into the form of model equations, and, further they are applied to determine the activity, property, and/or toxicity of compounds that are not tested experimentally or new structurally similar set of systems. Because mathematical models are utilized to develop equations, the quantification of different properties of a set of molecules is also readily carried out. Thus, the QSAR based models become useful toward predicting chemical activity and toxicity of molecules. They act as a novel technique to build models correlating structure, activity, and toxicity. QSAR-based studies have shown their applications in many fields like ecotoxicology, drug discovery, antitumor, molecular modeling, biotoxicity, chemico-biological interactions, toxicity predictions, gastrointestinal absorption, activity of peptides, pharmacokinetics/toxicokinetics, data mining, drug metabolism, determination of anti-HIV enzyme inhibitors, antitumor enzyme inhibitors, anticancer drugs, coinage metals, pesticide toxicity, fragrance, nanotoxicity, and in many other fields [9,18,29–37].

Apart from traditional QSAR methods, other QSAR techniques have also been developed. Three-dimensional correlation models (3D-QSAR) is another widely used method for analyzing the structure–property relationships. Using statistical correlation methods, their algorithm analyzes the relationship between the biological activity of a set of compounds and their 3D properties. Based on the applied energy functions involving force field calculations, it inspects both the steric fields (shape of the molecule) and the electrostatic fields. The other type of QSAR involves in silico tools like 3D Markovian Electron Delocalization Negentropies for toxicity predictions and 3D-QSAR comparative molecular field analysis (CoMFA) have also been applied extensively in the discovery and design of different molecules having desired properties [38]. CoMFA tries to predict the properties of a molecule by developing correlations between the structure and the pharmacological activity of compounds [39]. The potency of it in the design of effective enzyme inhibitors has also been explored. CoMFA has been applied for studying the toxicity of different organic polychlorinated derivatives with aryl-hydrocarbon receptors [40]. QSAR prediction strength lies in the information extracted from different databases. Some articles highlight various computational approaches for better correlative predictions, the role of QSAR in chemical toxicology, and prediction of biotoxicity [41]. Different topological descriptors [42] based QSAR models have been developed to correlate the hazardous effect of chemical compounds on the ecosystem [43].

DESCRIPTORS FOR PREDICTIVE TOXICOLOGY

A molecular descriptor is a structural or physicochemical property of a molecule or part of a molecule. A QSAR model for predictive toxicology is a mathematical relationship between a chemical's quantitative molecular descriptors and its toxicological endpoint [9,44]. Molecular descriptors derived from atomic or molecular properties that translate physicochemical, topological, and surface properties of compounds to establish the foundation for in silico predictive toxicology. Table 5.2 provides various molecular descriptors within the purview of structure–toxicity relationships have been proposed to predict toxicity properties of molecules. Chemoinformatics approaches including quantum chemical methods and molecular modeling techniques enable the definition of a large number of structural, molecular, and local quantities characterizing the reactivity, shape, and binding properties like atomic charges, molecular orbital energies, frontier orbital densities, superdelocalizabilities, atom–atom polarizabilities, molecular polarizability, charge transfer, dipole moment, and polarity indices and energy of a complete molecule as well as of molecular fragments and substituents [24,45,46]. These descriptors have been used to provide insights into the molecular effects that play an important role in a given chemical's toxicity, especially those aiding in constructing correlation models based on the observed manifestations [47]. Popular qualitative chemical concepts such as density functional theory (DFT) based electronegativity and hardness have been widely used in understanding various aspects of chemical reactivity [48]. The nature of basic chemical concepts, electronegativity, hardness, and softness, called global reactivity descriptors (GRD), has been theoretically justified within the framework of DFT. Along with these global descriptors, other important local reactivity descriptors (LRD), such as Fukui function and local softness, were also proposed to rationalize the reactivity of a particular atomic site or group in a molecule [24]. A series of new reactivity descriptors such as electrophilicity [49], local electrophilicity index [50], and group philicity [51] have been defined to understand the chemical reactivity and site selectivity. Generally GRD are used to probe the global reactivity of the molecules whereas the LRD provide information about the particular site in the molecule.

TABLE 5.2 Examples of Commonly Used Descriptors In Silico Toxicology to Construct Predictive Models

Descriptor Type/Group	Definition
Electronic descriptors	These represent diverse properties which are associated with many effects
E_{LUMO}	Energy of the lowest unoccupied molecular orbital
E_{HOMO}	Energy of the highest occupied molecular orbital
μ	Chemical potential (negative of electronegativity) $\mu = (E_{LUMO} + E_{HOMO})/2$
η	Chemical hardness $\eta = (E_{LUMO} - E_{HOMO})/2$
σ	Chemical softness $\sigma = 1 - \eta$
ω	Electrophilicity $\omega = \mu^2/2\eta$
A_{max}	Maximum atomic acceptor superdelocalizability within a molecule
D_{max}	Maximum atomic donor superdelocalizability within a molecule
A_N	Atomic acceptor superdelocalizability for atom N
D_N	Atomic donor superdelocalizability for atom N
$A_{orbital}$	One-orbital delocalizability associated with a given orbital (e.g., LUMO or LUMO+1)
HD/HA	Hydrogen bond donating/accepting ability.
Electronegativity (χ)	Ability of an atom (or group) to attract electrons associated with reactivity.
Atomic charge (q_n)	Charge associated with atom "n"
Dipole moment (δ)	Distribution of charge within a molecule
Steric descriptors	Associated with ability to reach target site (e.g., be absorbed across relevant biological membranes) and fit within specific receptors
Molecular weight	Relative molecular mass indicating general size of the molecule.
Molecular volume	This may be calculated using the sum of van der Waals atomic volumes; indicates general size.
Molecular surface area/solvent accessible surface area	Computationally, a probe molecule can be "rolled" over the surface of a molecule to determine the area that is accessible (to solvents or interacting molecules such as receptors)

K	The kappa index is a shape parameter based on the degree of branching of the molecular graph.
Sterimol (L_1, B_1–B_5)	Shape descriptors that indicate the length (L) of a substituent and its widths in different directions (B_1–B_5)
Es	The Taft steric constant indicates the size contribution of substituents on a parent molecule
Topological descriptors	
$^n\chi$	Nth order connectivity index
$^n\chi^v$	Valence corrected connectivity indices are used to distinguish between heteroatoms
3D descriptors	Three-dimensional representation of molecules provides a more accurate description of molecular dimensionality
Composite parameters	These represent combination effects and can provide additional information reflective of more than one feature
Polar surface area; hydrophobic surface area	Dividing the surface area of a molecule into regions of polarity or hydrophobicity can provide useful information for example in terms of specific receptor binding interactions
Electro topological state indices	A combination of electronic features and topological environment for given atoms
Functional groups/structural alerts	Certain toxicities may be associated with specific structural features
Similarity indices or similarity scores	Similarity of size, shape, spatial distribution of key atoms/functional groups, reactive potential
Indicator variables	These indicate the presence or absence (usually denoted by 1 or 0, respectively) of specific structural features (e.g., hydrogen bond donating/accepting groups, presence of particular functional group)
Hydrophobic/hydrophilic descriptors	Indicate solubility in aqueous and/or organic medium and relative partitioning between phases
Log P	Logarithm of the partition coefficient
Log D	Logarithm of the distribution coefficient
Log k'	Logarithm of the high performance liquid chromatography capacity factor
Log S_{aq}	Logarithm of the aqueous solubility
Lipole	Distribution of lipophilicity within a substituent or whole molecule
Molecular lipophilicity potential	Geometric distribution of lipophilicity within a molecule
π	Substituent constant indicating the influence of individual substituents on overall partitioning behavior: $\pi = \log P_{(substituted\ derivative)} - \log P_{(parent)}$

The applications of these descriptors toward the prediction of chemical reactivity, especially in the prediction of toxicity of polychlorinated biphenyls [52,53], polychlorinated dibenzofurans [54], and benzidines [55], biological activity of testosterone and estrogen derivatives [46] and other chemical reactive site and group identification studies have been of demonstrated to establish predictive toxicity models of potentially toxic molecules [56]. Considering an appropriate rationale to develop suitable predictive model depends on identifying key molecular descriptors of compounds for predicting activity. These descriptors are then correlated with a toxicological endpoints through appropriate statistical approach, such as linear multiple regression, discriminant analysis, recursive partitioning, or artificial neural networks [57,58].

DATABASES AND WEB TOOLS FOR PREDICTIVE TOXICOLOGY

To support the development and use of in silico methods, it is highly necessary to organize the chemical and toxicity data in a consistent way. Currently, there is a large number of toxicity data available, which are collected across several domains, and numerous research works have been published in the literature or online databases [43]. In silico toxicology models improve several applications in the alternative testing era to estimate potential toxic effects of a large number of chemicals using data on individual chemicals and their relationship to other series of chemicals/drugs [59]. Developed predictive approaches cannot completely replace standard bioassays or protocols, however, they can provide an earlier decision-making process until required data become available. Table 5.3 provides various in silico predictive toxicology data sources, online and standalone methods that are commonly employed for QSAR prediction and modeling. Some of the predictive toxicity models are publicly available as open web-based services for SAR modeling, molecular docking, QSAR predictive model for acute toxicity, antitarget activity, ecotoxicity, databases, and validated predictive models on the ADMET (absorption, distribution, metabolism, and excretion—toxicity) endpoints, blood–brain barrier permeability, human intestinal absorption, etc. [60,61] Computational tools for the visualization and analysis of the results are also available. In addition, researchers can use online/standalone computational tools for uploading their own data to develop models or to screen the compounds (toxicophores) using a variety of molecular descriptors and machine learning techniques [62]. Many commercial and open source software predictive toxicology tools (TOPKAT, CASE, and MultiCASE, Derek (Lhasa), and ToxTree) allow us to predict toxicity from the structure of chemical/drug compounds. Some in silico tools for toxicity prediction are simpler and other in silico tools are highly complex, involving multivariate modeling of large toxicological databases. This is an enormously promising area for the development of more sophisticated algorithms/methods/models to improve the predictive accuracy for improving risk assessment of chemicals. However, it should be recognized that the toxicity data/models are generated with very specific purposes and there is no unified

TABLE 5.3 Predictive Toxicology Relevant Data Sources: Databases, Online and Standalone Toxicity Predictive Tools[a]

Data Source	Description	Type
Pathguide	Contains information about 325 biological pathway related resources and molecular interaction related resources including Protein-Protein interactions and toxicity	A pathway resource list [http://www.pathguide.org/]
EYESOPEN	ADME (absorption, distribution, metabolism, and excretion/elimination/tox eyes open, commercial	ADME/tox, online [https://www.eyesopen.com/]
PharmGKB	A knowledge base that captures the relationships between drugs, diseases/phenotypes and genes involved in pharmacokinetics and pharmacodynamics	ADME/tox; PK/PD, online [https://www.pharmgkb.org/]
Myc	MycPermCheck: Online tool for permeability prediction of small molecules against *Mycobacterium tuberculosis*. Basis of the prediction program is a logistic regression model of the physico-chemical properties of permeable substances	Chemistry tools, online [http://www.mycpermcheck.aksotriffer.pharmazie.uni-wuerzburg.de/]
Edetox	Find compound properties	Chemistry, online [https://apps.ncl.ac.uk/edetox/]
LIVERTOX	Search for hepatoxicity of drugs and herbs	Database [https://livertox.nih.gov/]
Danish	Danish quantitative structure activity relationship (QSAR) database. A repository of estimates from over 70 (Q)SAR models for 166,072 chemicals	Database [https://eurl-ecvam.jrc.ec.europa.eu/]
eTOX	The eTOX Library of Public Resources for in silico toxicity prediction	Database [http://www.etoxproject.eu/]
ToxCast	Screening chemicals to predict toxicity faster and better	Database [https://www.epa.gov/chemical-research/toxicity-forecasting]
CEBS	Chemical effects in biological systems knowledge base (application of systems biology to ADME/Tox)	Database [https://cebs.niehs.nih.gov/]
PASS	PASS Inet predicts over 4000 kinds of biological activity, including pharmacological effects, mechanisms of action, toxic and adverse effects, interaction with metabolic enzymes and transporters, influence on gene expression, etc.	Online [www.pharmaexpert.ru/passonline/]
AERS spider	An online interactive tool to mine statistical associations in adverse event reporting system pharmacoepidemiology and drug safety	Online [http://www.chemoprofiling.org/AERS/]
SwissADME	Compute physicochemical descriptors as well as predict pharmacokinetics properties and drug like nature of one or multiple small molecules (BBB, Cyp, Pgp)	Online [www.swissadme.ch/]

Continued

TABLE 5.3 Predictive Toxicology Relevant Data Sources: Databases, Online and Standalone Toxicity Predictive Tools[a]—cont'd

Data Source	Description	Type
DILI	DILIserver: Deep Learning for Drug-Induced Liver Injury, this has been the single most frequent cause of safety-related drug marketing withdrawals for the past 50years	Online [http://www.pkumdll.cn/DILIserver/DILIhome.php]
Open	OpenVirtualToxLab: ADMET prediction, 16 proteins, known or suspected to trigger adverse effects are implemented at present: 10 nuclear receptors (androgen, estrogen a, estrogen b, glucocorticoid, liver X, mineralocorticoid, peroxisome proliferator-activated receptor g, progesterone, thyroid a, thyroid b), four members of the cytochrome P450 enzyme family (1A2, 2C9, 2D6, 3A4), a cytosolic transcription factor (aryl hydrocarbon receptor) and a potassium ion channel (hERG).	Online [http://www.biograf.ch/index.php?id=home]
iPrior	Using online tool for modeling ToxCast-TM assays towards prioritization of animal toxicity testing	Online [[http://iprior.ochem.eu]]
Chemo tools	Fraggle, the fragment store (search for fragments by drug names or PDB header). It provides property information (charge, hydrophobicity, and binding site preferences) and performs statistical analysis and can view the IDS of drugs and toxic compounds, which contain the fragments	Online [http://bioinf-applied.charite.de/fragment_store/]
SCYPPred	A web-based predictor of SNPs for human cytochrome P450	Online [https://omictools.com/scyppred-tool]
AOP	Define adverse outcome pathways (AOPs)	Online [http://aopkb.org/]
AOP-XPlorer	Explore AOPs	Online [aopxplorer.org]
Effectopedia	It is an open-knowledge aggregation and collaboration tool designed to facilitate the interdisciplinary efforts for delineating AOPs	Online [https://www.effectopedia.org/]
IntSide	A web server for the chemical and biological examination of drug side effects	Online [https://omictools.com/intside-tool]
ToxPredict	ToxPredict (associated to OpenTox)	Online [https://apps.ideaconsult.net/ToxPredict]
admetSAR	Models and databases: provides the manually curated data for diverse chemicals associated with known absorption, distribution, metabolism, excretion, and toxicity profiles. many endpoints, hERG, CYP, PgP, oral, toxicity, Ames test	Online [http://lmmd.ecust.edu.cn:8000/]
hERG	The hERG 1.0 server predicts cardiotoxicity of drug molecules	Online [http://www.cbs.dtu.dk/CBS/services/hERG/]

Name	Description	Availability
ADME SARfari	Predict likely ADME targets for an input molecule, Find ADME targets similar to an input FASTA sequence, Find ADME targets related to text terms, Find pharmacokinetic data relating to an input target, sequence or text term, Find activity pharmacokinetic data for an input molecule or related compounds (via a similarity substructure search), match expression levels in human tissues for found targets	Online [https://www.ebi.ac.uk/chembl/admesarfari]
ProTox	A web server for the in silico prediction of rodent oral toxicity	Online [http://tox.charite.de/]
Alkemio	Association of chemicals with biomedical topics by text and data mining	Online [http://cbdm.mdc-berlin.de/~medlineranker/cms/alkemio]
ACD/I-Lab	ADME-Tox prediction	Online [https://ilab.acdlabs.com/]
ToxCreate	Creates computational models to predict toxicity	Online [http://www.toxcreate.net/]
HExpoChem	The server contains information on diverse sources of chemicals with the aim to explore human health risk from diverse chemicals exposure. Five sources of information are considered i.e., drugs, foods, cosmetics, industrial chemicals and human metabolites corresponding of 10,183 unique chemicals with bioactivities for 19,483 human proteins. HExpoChem can help in the decision of potential proteins and proteins complexes associated to life style diseases. It can assist also to the possible cumulative risk of chemicals that interact to a same set of proteins	Online [http://www.cbs.dtu.dk/services/HExpoChem-1.0/]
QED	A webserver for quantitative estimating the drug likeness of a molecule	Online [http://crdd.osdd.net/oscadd/qed/]
DrugMint	Predict the druggabilty of a compound. This server has been developed on the basis of difference in descriptors of approved and experimental small molecules. This server will help in knowing the druggable properties of a chemical structure	Online [http://crdd.osdd.net/oscadd/drugmint/]
Mcule	Toxicity checker, searching for substructures commonly found in toxic and promiscuous ligands, based on more than 100 SMARTS toxic matching rules	Online [https://mcule.com/]
Gusar	Gusar ecotoxicity	Online [http://www.way2drug.com/gusar/environmental.html]
Gusar	Gusar antitargets, rat toxicity	Online [http://www.way2drug.com/gusar/antitargets.html]
CoFFer	A QSAR web service that predicts chemical compounds and provides fragments to aid interpreting predictions. QSAR models (ADMET, off targets, repositioning)	Online [http://coffer.informatik.uni-mainz.de/]
Gusar	Gusar acute rat toxicity prediction	Online [http://www.way2drug.com/gusar/acutoxpredict.html]

Continued

TABLE 5.3 Predictive Toxicology Relevant Data Sources: Databases, Online and Standalone Toxicity Predictive Tools[a]—cont'd

Data Source	Description	Type
ToxiPred	Prediction of aqueous toxicity of small chemical molecules in *T. pyriformis*	Online [crdd.osdd.net/raghava/toxipred]
Lazar	Lazy structure–activity relationships is a tool for the prediction of toxic activities	Online [http://lazar.in-silico.de/]
VirtualToxLab	The VirtualToxLab, 3D ADME/tox	Online [https://omictools.com/virtualtoxlab-tool]
Toxpredict	Web service to estimate toxicological hazard of a chemical structure. Molecules can be drawn, or input by any identifier (CAS, Name, EINECS) or SMILES or InChI or URL of OpenTox compound or dataset	Online [https://apps.ideaconsult.net/ToxPredict]
AlogP	Tools to predict logP (with several methods)	Online [www.vcclab.org/web/alogps/]
ZINC	Some ADME/tox filtering	Online [http://zinc.docking.org/]
ICM	ADME/tox molsoft, commercial and demo	Online [http://www.molsoft.com/mprop]
Chemaxon	ADME/tox chemical	Online [https://chemicalize.com/welcome]
MOLNetwork	ADME/tox online molecular networks	Online [https://github.com/pyeguy/MolNetwork]
ToxAlerts	Web server of structural alerts for toxic chemicals and compounds with potential adverse reactions. The database already contains almost 600 structural alerts for such endpoints as mutagenicity, carcinogenicity, skin sensitization, compounds that undergo metabolic activation, and compounds that form reactive metabolites and, thus, can cause adverse reactions, QSAR models can be built	Online [https://omictools.com/toxalerts-tool]
EPA USA	TEST estimates the toxicity values and physical properties of organic chemicals based on the molecular structure of the organic chemical entered by the user	Standalone [https://www.epa.gov/chemical-research/toxicity-estimation-software-tool-test]
Tox-Comp.net	Prediction for hERG	Standalone [http://www.tox-comp.net/]
Vega-QSAR	Vega-QSAR.eu: Tools for QSAR prediction of ADMET (ORCHESTRA is an EU project, funded to disseminate recent research on in silico (computer-based) methods for evaluating the toxicity of chemicals, REACH, courses); ANTARES, alternative to animal testing, CAESAR, EC funded project which was specifically dedicated to develop QSAR models for the REACH legislation	Standalone [http://www.vega-qsar.eu/]

Name	Description	Availability
Natural Product Likeness	The Natural-Product-Likeness scoring system is also implemented as workflows, and is available under Creative Commons Attribution-Share Alike 3.0. The present link goes to the executable standalone java package (under Academic Free License)	Standalone [http://sourceforge.net/projects/np-likeness/]
PaDEL	PaDEL-DDPredictor: Calculate pharmacodynamics, pharmacokinetics and toxicological properties of compounds	Standalone [http://www.yapcwsoft.com/dd/padelddpredictor/]
QSAR TOOLBOX	Software for grouping chemicals into categories and filling gaps in (eco)toxicity data needed for assessing the hazard of chemicals	Standalone [http://oasis-lmc.org/products/software/toolbox.aspx]
Toxtree	Application for grouping chemicals and for predicting various types of toxicity based on decision tree approaches.	Standalone [http://lhcp.jrc.ec.europa.eu/our_labs/eurl-ecvam/laboratories-research/predictive_toxicology]
Checkmol	Checkmol is a command-line utility program which reads molecular structure files in different formats	Standalone [http://merian.pch.univie.ac.at/~nhaider/cheminf/cmmm.html]
Lilly Open Innovation ADMET	Rejection rules from Lilly Open Innovation Drug Discovery initiative (reactivity and promiscuity filters, drug similarity). Over 275 rules addressing a variety of possible reasons to reject a molecules and the BMS rules. Lilly-Med Chem-Rules stand-alone distribution. Thanks to Greg Durst for pointing to the Lilly-Med Chem-Rules stand-alone command line utility titled "tsubstructure", which lets you search very large SMILES files for specific SMARTS queries	Standalone [https://github.com/IanAWatson/Lilly-Medchem-Rules]
MetaSite	Predict the site of metabolism for substrates of 2C9, 2D6, 3A4, 1A2, and 2C19 cytochromes	Standalone, commercial [http://www.moldiscovery.com/soft_metasite.php]
MedChem Designer	MedChem Designer 2.0, chemical drawing and ADMET prediction	Standalone [https://simplus-downloads.com/index2.html]
TissueTool	Tissue distribution databases: A repository of tissue distribution profiles for identifying and ranking the genes in the spectrum of tissue specificity based on expressed sequence tags. This repository is currently available for several model organisms across animal and plant kingdoms and is fundamentally based on the UniGene database	Tissue distribution of several targets, online [http://genome.dkfz-heidelberg.de/menu/tissue_db/examples.html]

*As on April 2017.
BBB, blood–brain barrier; BMS, Bristol-Myers Squibb; hERG, human Ether-à-go-go Related Gene; IDS, identifications; PDB, Protein Data Bank; SNPs, single-nucleotide polymorphisms; TEST, Toxicity Estimation Software Tool.

in silico platform for toxicity prediction. It is essential to ensure the validity of any predictive toxicity model, nature/use of chemical for predictive applications, choice of descriptors, rationalization of the process and approach, and relationship with available toxicity endpoints should be critically considered in a context-dependent pragmatic way.

SUMMARY

Determining the toxicological effects of chemicals as early as possible is essential for human health and the environmental safety. Toxicity of a chemical/drug refers to the adverse effect on the whole organism (animal), particular organ (liver), or substructure of the organism (cell). However, in vivo and in vitro high-throughput assays are expensive and time consuming. In silico toxicity predictive models offer robust and economical alternative to in vivo and in vitro bioassays and reduce animal experiments as well as experimental resources. The main objective of this chapter is to summarize various aspects of in silico models used to solve many problems in predictive toxicology and furthering knowledge on toxicity mechanisms to design safe chemicals/drugs/materials. This chapter provides a framework for the development and use of in silico models. There are a large number of in silico models are developed to categorize the toxicity of chemical compounds, the toxicological endpoints or the effect of different concentrations of the chemicals. *Structure–activity–toxicity* models for predicting activity of compounds can be developed based on knowledge of chemical structure and particular properties or relevant combination of descriptor series. The properties such as physicochemical, structural, and electronic properties can be computed using a range of software suites and correlated with the information determined experimentally. A variety of QSAR approaches are applied in diverse areas covering a range of endpoints (LD_{50}, LC_{50}, EC_{50}, etc.) to predict the biological activities, pharmacokinetic properties, and toxicities, as well as physicochemical properties of drugs and drug-like compounds. This chapter also covered toxicity databases and online service tools used for predicting toxicity of different types of substances both quantitatively and qualitatively. An increasing number of open toxicity in silico prediction sources are available and some of them are commercial. Implementing and integrating in silico methods into regulatory risk assessment is rapidly evolving. In silico predictive toxicology enables toxicity regulation/safety formulation to impact the discovery of chemicals/drugs with a superior safety profile. Continuous development of more physiologically relevant predictive toxicity model systems, advanced algorithms, and analysis methods can replace battery of in vitro and in vivo toxicity tests in the near future.

ACKNOWLEDGMENTS

Funding from the Council of Scientific & Industrial Research (CSIR), New Delhi and CSIR-Indian Institute of Toxicology Research, Lucknow is gratefully acknowledged.

REFERENCES

[1] Penningroth S. Essentials of toxic chemical risk: science and society. CRC Press; 2016.

[2] Council NR. Toxicity testing in the 21st century: a vision and a strategy. National Academies Press; 2007.

[3] Raunio H. In silico toxicology – non-testing methods. Front Pharmacol 2011;2:33.

[4] Krewski D, Acosta D, Andersen M, Anderson H, Bailar JC, Boekelheide K, et al. Toxicity testing in the 21st century: a vision and a strategy. J Toxicol Environ Health B Crit Rev 2010;13(0):51–138.

[5] Parasuraman S. Toxicological screening. J Pharmacol Pharmacother 2011;2(2):74–9.

[6] Rovida C, Asakura S, Daneshian M, Hofman-Huether H, Leist M, Meunier L, et al. Toxicity testing in the 21st century beyond environmental chemicals. Altern Animal Exp ALTEX 2015;32(3):171–81.

[7] Hartung T, Luechtefeld T, Maertens A, Kleensang A. Food for thought… integrated testing strategies for safety assessments. Altex 2013;30(1):3.

[8] Hamm J, Sullivan K, Clippinger AJ, Strickland J, Bell S, Bhhatarai B, et al. Alternative approaches for identifying acute systemic toxicity: moving from research to regulatory testing. Toxicol In Vitro 2017.

[9] Raies AB, Bajic VB. In silico toxicology: computational methods for the prediction of chemical toxicity. Wiley Interdiscip Rev Comput Mol Sci 2016;6(2):147–72.

[10] Cronin MTD, Madden JC. In silico toxicology-an introduction. In Silico Toxicology: Principles and ApplicationsThe Royal Society of Chemistry; 2010. p. 1–10. [Chapter 1].

[11] Benigni R, Battistelli CL, Bossa C, Colafranceschi M, Tcheremenskaia O. Mutagenicity, carcinogenicity, and other end points. Comput Toxicol 2013;II:67–98.

[12] Higgins TE, Sachdev JA, Engleman SA. Toxic chemicals: risk prevention through use reduction. CRC Press; 2016.

[13] Wilson AGE. Introduction and overview. New Horizons in predictive toxicology: current Status and applicationThe Royal Society of Chemistry; 2012. p. 1–8. [Chapter 1].

[14] Doke SK, Dhawale SC. Alternatives to animal testing: a review. Saudi Pharm J 2015;23(3):223–9.

[15] Ranganatha N, Kuppast I. A review on alternatives to animal testing methods in drug development. Int J Pharm Pharm Sci 2012;4(Suppl. 5):28–32.

[16] Cronin MTD. In silico tools for toxicity prediction. New Horizons in predictive toxicology: current Status and applicationThe Royal Society of Chemistry; 2012. p. 9–25. [Chapter 2].

[17] Sliwoski G, Kothiwale S, Meiler J, Lowe EW. Computational methods in drug discovery. Pharmacol Rev 2014;66(1):334–95.

[18] Knudsen TB, Keller DA, Sander M, Carney EW, Doerrer NG, Eaton DL, et al. FutureTox II: in vitro data and in silico models for predictive toxicology. Toxicol Sci 2015;143(2):256–67.

[19] Valerio LG. In silico toxicology for the pharmaceutical sciences. Toxicol Applied Pharmacology 2009;241(3):356–70.

[20] Wilk-Zasadna I, Bernasconi C, Pelkonen O, Coecke S. Biotransformation in vitro: an essential consideration in the quantitative in vitro-to-in vivo extrapolation (QIVIVE) of toxicity data. Toxicology 2015;332:8–19.

[21] Groh KJ, Carvalho RN, Chipman JK, Denslow ND, Halder M, Murphy CA, et al. Development and application of the adverse outcome pathway framework for understanding and predicting chronic toxicity: I. Challenges and research needs in ecotoxicology. Chemosphere 2015;120:764–77.

[22] Sturla SJ, Boobis AR, FitzGerald RE, Hoeng J, Kavlock RJ, Schirmer K, et al. Systems toxicology: from basic research to risk assessment. Chem Res Toxicol 2014;27(3):314–29.

[23] Burton J, Worth AP, Tsakovska I, Diukendjieva A. In silico models for acute systemic toxicity. In silico methods for predicting drug toxicity2016. p. 177–200.

[24] Parthasarathi R, Elango M, Padmanabhan J, Subramanian V, Roy D, Sarkar U, et al. Application of quantum chemical descriptors in computational medicinal chemistry and chemoinformatics. 2006.

[25] Hansch C, Maloney PP, Fujita T, Muir RM. Correlation of biological activity of phenoxyacetic acids with Hammett substituent constants and partition coefficients. Nature 1962;194(4824):178–80.

[26] Hansch C, Hoekman D, Gao H. Comparative QSAR: toward a deeper understanding of chemicobiological interactions. Chem Rev 1996;96(3):1045–76.

[27] Karelson M, Lobanov VS, Katritzky AR. Quantum-chemical descriptors in QSAR/QSPR studies. Chem Rev 1996;96(3):1027–44.

[28] Hansch C, Hoekman D, Leo A, Weininger D, Selassie CD. Chem-Bioinformatics: comparative QSAR at the interface between chemistry and biology. Chem Rev 2002;102(3):783–812.

[29] Cronin MT, Walker JD, Jaworska JS, Comber MH, Watts CD, Worth AP. Use of QSARs in international decision-making frameworks to predict ecologic effects and environmental fate of chemical substances. Environ Health Perspect 2003;111(10):1376.

[30] Cronin MT, Jaworska JS, Walker JD, Comber MH, Watts CD, Worth AP. Use of QSARs in international decision-making frameworks to predict health effects of chemical substances. Environ Health Perspect 2003;111(10):1391.

[31] Fourches D, Pu D, Tassa C, Weissleder R, Shaw SY, Mumper RJ, et al. Quantitative nanostructure-activity relationship (QNAR) modeling. ACS Nano 2010;4(10):5703–12.

[32] Puzyn T, Rasulev B, Gajewicz A, Hu X, Dasari TP, Michalkova A, et al. Using nano-QSAR to predict the cytotoxicity of metal oxide nanoparticles. Nat Nano 2011;6(3):175–8.

[33] Sjogren E, Thorn H, Tannergren C. In Silico modeling of gastrointestinal drug absorption: predictive performance of three physiologically based absorption models. Mol Pharmaceut 2016;13(6):1763–78.

[34] Madden J. Introduction to QSAR and other in silico methods to predict toxicity. In Silico Toxicol 2010:11–30.

[35] Selassie C, Verma RP. History of quantitative structure–activity relationships. Burger Med Chem Drug Discov 2003.

[36] Roncaglioni A, Benfenati E. In silico-aided prediction of biological properties of chemicals: oestrogen receptor-mediated effects. Chem Soc Rev 2008;37(3):441–50.

[37] Kumar A, Dhawan A, Shanker R. The need for novel approaches in ecotoxicity of engineered nanomaterials. J Biomed Nanotechnol 2011;7(1):79–80.

[38] Pan S, Gupta A, Roy DR, Sharma RK, Subramanian V, Mitra A, et al. Application of conceptual density functional theory in developing QSAR models and their usefulness in the prediction of biological activity and toxicity of molecules. Chemometrics applications and research: QSAR in Medicinal chemistryApple Academic Press; 2016. p. 183–214.

[39] Chakraborty A, Pan S, Chattaraj PK. Biological activity and toxicity: a conceptual DFT approach. Applications of density functional theory to biological and Bioinorganic chemistrySpringer; 2013. p. 143–79.

[40] Šala M, Balaban AT, Veber M, Pompea M. QSAR models for estimating aryl hydrocarbon receptor binding affinity of polychlorobiphenyls, polychlorodibenzodioxins, and polychlorodibenzofurans. Match Commun Math Comput Chem 2016;75(3):559–82.

[41] Eduati F, Mangravite LM, Wang T, Tang H, Bare JC, Huang R, et al. Prediction of human population responses to toxic compounds by a collaborative competition. Nat Biotech 2015;33(9):933–40.

[42] Randić M. Generalized molecular descriptors. J Math Chem 1991;7(1):155–68.

[43] Cherkasov A, Muratov EN, Fourches D, Varnek A, Baskin II, Cronin M, et al. QSAR Modeling: where have you been? Where are you going to? J Med Chem 2014;57(12):4977–5010.

[44] Guha R, Willighagen E. A survey of quantitative descriptions of molecular structure. Curr Topics Med Chem 2012;12(18):1946–56.

[45] Padmanabhan J, Parthasarathi R, Subramanian V, Chattaraj P. Electrophilicity-based charge transfer descriptor. J Phys Chem A 2007;111(7):1358–61.

[46] Parthasarathi R, Subramanian V, Roy D, Chattaraj P. Electrophilicity index as a possible descriptor of biological activity. Bioorg Med Chem 2004;12(21):5533–43.

[47] Roy D, Parthasarathi R, Maiti B, Subramanian V, Chattaraj P. Electrophilicity as a possible descriptor for toxicity prediction. Bioorg Med Chem 2005;13(10):3405–12.

[48] Geerlings P, De Proft F, Langenaeker W. Conceptual density functional theory. Chem Rev 2003;103(5):1793–874.

[49] Chattaraj PK, Sarkar U, Elango M, Parthasarathi R, Subramanian V. Electrophilicity as a possible descriptor of the kinetic behaviour. 2005. arXiv preprint physics/0509089.

[50] Chattaraj PK, Maiti B, Sarkar U. Philicity: a unified treatment of chemical reactivity and selectivity. J Phys Chem A 2003;107(25):4973–5.

[51] Parthasarathi R, Padmanabhan J, Elango M, Subramanian V, Chattaraj P. Intermolecular reactivity through the generalized philicity concept. Chem Phys Lett 2004;394(4–6):225–30.

[52] Parthasarathi R, Padmanabhan J, Subramanian V, Maiti B, Chattaraj P. Chemical reactivity profiles of two selected polychlorinated biphenyls. J Phys Chem A 2003;107(48):10346–52.

[53] Padmanabhan J, Parthasarathi R, Subramanian V, Chattaraj P. QSPR models for polychlorinated biphenyls: n-Octanol/water partition coefficient. Bioorg Med Chem 2006;14(4):1021–8.

[54] Sarkar U, Padmanabhan J, Parthasarathi R, Subramanian V, Chattaraj P. Toxicity analysis of polychlorinated dibenzofurans through global and local electrophilicities. J Mol Struct THEOCHEM 2006;758(2):119–25.

[55] Parthasarathi R, Padmanabhan J, Sarkar U, Maiti B, Subramanian V, Chattaraj PK. Toxicity analysis of benzidine through chemical reactivity and selectivity profiles: a DFT approach. Internet Electron J Mol Des 2003;2(12):798–813.

[56] Roy D, Sarkar U, Chattaraj P, Mitra A, Padmanabhan J, Parthasarathi R, et al. Analyzing toxicity through electrophilicity. Mol Diversity 2006;10(2):119–31.

[57] Gleeson MP, Modi S, Bender A, Marchese Robinson RL, Kirchmair J, Promkatkaew M, et al. The challenges involved in modeling toxicity data in silico: a review. Curr Pharmaceut Design 2012;18(9):1266–91.

[58] Carrió P, Sanz F, Pastor M. Toward a unifying strategy for the structure-based prediction of toxicological endpoints. Archives Toxicol 2016;90(10):2445–60.

[59] Só J, Jørgensen FS, Brunak S. Prediction methods and databases within chemoinformatics: emphasis on drugs and drug candidates. Bioinformatics 2005;21(10):2145–60.

[60] http://www.vls3d.com/links/chemoinformatics/admet/admet-and-physchem-predictions-and-related-tools.

[61] https://www.click2drug.org/directory_ADMET.html.

[62] Sushko I, Salmina E, Potemkin VA, Poda G, Tetko IV. ToxAlerts: a web server of structural alerts for toxic chemicals and compounds with potential adverse reactions. J Chem Inf Model 2012;52(8):2310–6.

Chapter 6

The Use of Transcriptional Profiling in In Vitro Systems to Determine the Potential Estrogenic Activity of Chemicals of Interest

Jorge M. Naciff

The Procter and Gamble Company, Mason, OH, United States

INTRODUCTION

One of the greatest challenges that the field of toxicology is currently facing is the development of reliable in vitro alternatives to animal testing to evaluate the safety of chemicals of interest for human use, particularly to address the potential for systemic toxicity [1,2]. Thus far, the accepted methodology to assess the potential for a chemical to elicit target organ-specific or systemic toxicity relies on the use of animal models in which the outcome of chemical exposure is the production of adverse effects at relatively high doses, assessed as changes in clinical pathology and histopathology compared to untreated controls, that are extrapolated to predict the adverse effects in humans that could result from the exposure to the same chemical. Thus, the development of in vitro assays to assess chemical safety, as alternatives to animal testing, has become an important research mission in toxicology. Moreover, the field has been moving more into the definition of adverse outcome pathways (AOPs) activated by chemical exposure and the use of this information to determine chemical safety, with the ultimate goal of relying less in animal derived-apical endpoints for chemical risk assessment. An AOP is a concept that provides a framework for organizing knowledge about the progression of toxicity events across scales of biological organization that lead to adverse outcomes relevant for risk assessment [3]. An AOP links molecular initiating events triggered by chemical exposures to consequent cellular key events that lead to adverse outcomes in an organism or a population [4]. Thus, the need is not only for in vitro predictive toxicological

In Vitro Toxicology. http://dx.doi.org/10.1016/B978-0-12-804667-8.00006-7

assays, but also for these assays to provide a significant assessment of the biological activity associated with the exposure to chemicals of interest, and with that, a better understanding of the underlying mechanisms of potential toxicity.

Exogenous chemicals have a plethora of endogenous molecular targets, but one common early event after chemical insult is the initiation of changes in gene expression in affected cells. The changes in gene expression may be directly involved in the signal transduction process (e.g., for steroid receptors) or an indirect response to cell damage or other changes. Therefore, analysis of gene expression is useful in identifying the earliest events in AOPs. Currently, one of the most advanced technologies, which can provide a great insight on the biological activity associated with chemical exposure is transcriptional profiling (transcriptomics), the evaluation of the transcriptional changes (messenger RNA [mRNA]) resulting from such exposure. The use of an appropriate in vitro model for chemical assessment, coupled with comprehensive transcriptional profiling and the proper bioinformatics tools, can help to define the biological effects of chemical exposure, at the molecular level. This approach can provide information to determine the relevant biological activities associated with specific chemical exposure and to define key molecular events that could elicit adverse (toxic) outcomes.

This chapter is focused in the use of transcriptional profiling in in vitro systems to determine the potential estrogenic activity of chemicals of interest, as an example of the use of in vitro systems to assess the biological activity associated with chemical exposure. However, this approach can be applied to address the potential effects of any given chemical in other hormonal systems, particularly for those chemicals with suspected endocrine disruptive activity, as well as other modes of action (MOAs), not mediated by a direct interaction between a ligand and a specific receptor, but that also involve changes in gene expression (see [5,6]).

WHY DO WE HAVE TO ASSESS THE POTENTIAL ESTROGENICITY OF CHEMICALS OF INTEREST?

In mammals, the estrogen system has important physiological roles from development to adulthood, in both females and males. This hormonal system is able to regulate the functions of multiple cell types and organs, from the reproductive organs to the central nervous system, regulating physiology and behavior. Disruption of the normal function of the estrogen system can result in significant adverse events to both humans and ecological species. Further, there are thousands of natural and manmade chemicals to which humans and wildlife are exposed via air, water, food, soil, workplace, and home [7–9], for which there is limited (at best) or no data (most cases) to determine their potential to disrupt the estrogen system.

The most potent form of endogenous mammalian estrogenic steroids is 17β-estradiol, an aromatized C18 steroid with hydroxyl groups at 3-beta and 17-beta position. In humans, it is produced primarily by the ovaries and the placenta. It is also produced by the arterial wall (from the aromatase-mediated conversion of testosterone) as well as other tissues, including the skin, bone, brain, the vascular endothelium, and aortic smooth muscle cells, and adipose tissue of men and postmenopausal women [10,11].

One of the most familiar physiological functions of the estrogen system is the regulation of the development, differentiation, and maintenance of female reproductive system, and not just the uterus and mammary glands, but also tissues and organs responsible for the secondary sex characteristics. At maturity, when the uterus and ovaries undergo the highly coordinated cycles of proliferation and differentiation associated with estrous cycle, the activity of estrogen system is essential. Estrogens and other extracellular signals coordinate the estrous cycle through the tight regulation of key signaling molecules. However, estrogen also regulates other physiological systems of the body [12] and not only in females but also in males. For example, in men the estrogen system is essential for modulating erectile function, spermatogenesis, and libido [13]. Functions for estrogen in the endocrine, cardiovascular, nervous, musculoskeletal and the immune system, and metabolism are all well documented [12–16].

The profound effects of 17β-estradiol on cell growth, differentiation, and general homeostasis of the reproductive and other systems are mediated mainly by the regulation of temporal and cell type-specific expression of different genes, whose products are the molecules controlling those physiological events regulated by estrogen [17–19]. The expression of those genes could be directly regulated by either of the two estrogen receptor (ER) isoforms (ERα or ERβ), which form homo- or heterodimers that translocate to the nucleus and regulate expression of ER target genes by directly binding an estrogen response element, functioning as ligand-dependent transcription factors, or forming protein–protein interactions to indirectly regulate other transcription factors that act on the targeted genes [17–19]. Transcriptional regulation can also be accomplished by estrogen through alternative pathways requiring the participation specific regulatory proteins, such as AP-1 [20]; or indirectly via one of the multiple gene products induced by estrogen, through the regulation of regulatory factors such as insulin-like growth factor 1, transforming growth factor beta (TGFβ), etc., or even by actions of estrogen modifying the activity of other transcriptional regulators, such as nuclear factor kappa B, which mediates specific gene expression changes [21,22]. Although both ERα or ERβ can mediate rapid signaling outside of the nucleus, throughout protein–protein interactions of these receptors with adaptor molecules, G proteins and kinases [23], the G protein-coupled estrogen receptor (GPER) (or GPR30), the third ER in mammals, is a critical mediator of rapid signaling in response to estrogen that is independent of transcriptional changes [24]. GPER's extranuclear actions are mainly mediated by eliciting changes in cyclic adenosine monophosphate production and intracellular Ca^{2+} levels [25]. However, GPER also mediates estrogen's transcriptional regulation either directly as a transcription factor, or indirectly via cyclic adenosine monophosphate or intracellular Ca^{2+} concentration changes [26,27].

The specificity and sensitivity of the ERs for 17 β-estradiol and other endogenous ligands are very high (dissociation constant, Kd, in the sub-to-nM range; [28]). However, there are natural and synthetic chemicals capable of interacting and modifying the ERs' activity and thus modify the physiological response of this hormonal system in humans and animals, in some cases activating this system and in others inactivating it. Given the multiplicity of physiological functions regulated by estrogen and its receptors, and the possibility of these functions to be affected by exogenous chemicals capable of modifying these functions, it is critical to assess the estrogenic (or antiestrogenic) potential of chemicals placed in commerce with the potential to result in significant environmental exposure, as part of their risk assessment.

A multiagency effort called toxicology in the 21st century (Tox21), which includes scientists from the National Institute of Environmental Health Sciences—National Toxicology Program, the National Institutes of Health's (NIH's) National Center for Advancing Translational Sciences, the Environmental Protection Agency's (EPA's) National Center for Computational Toxicology, and the Food and Drug Administration, is aimed at developing new technology, using in vitro high-throughput screening (HTS), and computational methods, that will further speedup chemical screening for all types of activity, increase accuracy of the results, and reduce the use of laboratory animals [29].

The objective of the Tox21 partnership is to shift the assessment of chemical hazards away from traditional laboratory animal toxicology studies to an evaluation based on target-specific, mechanism-based, biologic observations largely obtained by using in vitro assays, with the ultimate aim of improving risk assessment for humans and the environment. The Tox21's strategy is to use in vitro HTS and high-content screening technologies to test a broad variety of chemicals and consider data from those screens collectively to assess effects on biologic pathways related to toxicity. The purpose is that the data from Tox21 testing will be used to develop a better understanding of these toxicity or AOPs, ideally enabling the eventual use of in vitro assay data to predict the adverse effects of chemical exposures in vivo (http://ntp.niehs.nih.gov/go/tox21). As part of the Tox21 efforts, EPA's Toxicity Forecaster (ToxCast) has generated data on over 1800 chemicals from a broad range of sources including industrial and consumer products, food additives, and other chemicals of interest to EPA. These chemicals have been screened in over 800 high-throughput assays that cover a range of high-level cell responses and approximately 300 signaling pathways (https://www.epa.gov/chemical-research/toxicity-forecasting). One of these pathways addresses the estrogen system, by using the data from 18 in vitro assays, which include biochemical and cell-based in vitro assays that probe perturbations of ER pathway responses at sites within the cell: receptor binding, receptor dimerization, chromatin binding of the mature transcription factor, gene transcription, and changes in ER-induced cell growth kinetics [30]. The data from these 18 ER ToxCast HTS assays have been integrated into a computational model that can discriminate bioactivity from assay-specific interference and cytotoxicity [31]. However, this model can only discriminate chemical's potential ER agonist bioactivity, and mostly toward the ERα. In a more specific assessment, and still under the Tox21 effort, a library of approximately 10,000 environmental chemicals and drugs have been screened by EPA's scientist in three independent runs for ERα agonist and antagonist activity using two types of ER reporter gene cell lines, one with an endogenous full length ERα (ER-luc; BG1 cell line) and the other with a transfected partial receptor consisting of the ligand binding domain (ER-bla; ERα β-lactamase cell line), in a quantitative HTS (qHTS) format [32]. The authors concluded that qHTS is a feasible option to identify environmental chemicals with the potential to interact with the ERα signaling pathway. Continuing with the Tox21 program, this library of 10,000 compounds has been screened (using in vitro HTS approaches) for its activity in other receptors: including androgen receptor, retinoic acid receptor, and other nuclear receptors, searching simultaneously for agonists and antagonists in a pairwise manner. Data generated by these projects are being deposited in PubChem (https://pubchem.ncbi.nlm.nih.gov/) from The National Center for Biotechnology Information.

Although EPA's work is relevant to better understand the effect of chemical exposure on the estrogen system, as well as other systems and pathways, the assays they have used may not detect chemicals that are ERβ-selective or act through nonclassical ER mechanisms (nongenomic), or through GPER. To address these shortcomings, a new set of assays need to be developed or a completely new approach should be taken to cover them. A transcriptional profiling approach in the appropriate biological system (i.e., responsive to estrogen exposure) will be definitively more comprehensive and result in the identification of key gene expression changes associated with the disruption of the estrogen system through any of the ERs.

Transcriptional Profiling in In Vitro Systems to Determine the Potential Estrogenic Activity of Chemicals of Interest

The technological advancements that made possible the sequencing of the human genome, and that of many other organisms, have been accompanied by the development of technologies that allow one to evaluate mRNA or microRNA expression in a comprehensive manner. One of the most robust approaches currently being used to evaluate transcriptional changes involves the use of comprehensive microarrays, which allow for genome-wide expression profiling and evaluation of the whole-transcriptome in any given biological sample. With these microarrays, it is possible to evaluate changes in the expression of every gene (evaluating the mRNA and long intergenic noncoding RNA transcripts containing important regulatory elements) in a biological sample from difference organisms, being cells in culture or tissues from in vivo experiments, in response to an external perturbation, including those elicited by chemical exposure [33]. Whole transcriptome sequencing (RNA-seq) is an emerging approach also used for comprehensive transcriptional profiling [34], however, its use to assess chemical exposure's biological outcomes is still in its early stage.

Comprehensive transcriptional profiling of a biological sample collected after chemical exposure can provide the fundamental information to define the chemical's biological activity needed to identify the principal mode(s) of action for any given chemical (e.g., ER regulation) and with that its potential toxicity. At the same time, gene expression data can be used to establish the quantitative relationships between chemical exposure and effects on discrete biological pathways linked to dose– and time–response relationships, a requirement to identify molecular initiating events, as well as key biological events associated with an adverse outcome, information necessary to define an AOP associated with such chemical exposure.

As indicated, chemicals capable of modifying the normal function of the estrogen system, directly or indirectly modify the expression of specific genes. The identification of these gene expression changes, and their encoded products, is essential to define critical biological effects of these chemicals, which are part of the initial molecular event and as such could be conductive to activation of a specific AOP. The same is true for other chemicals capable of modifying any of the hormonal systems, namely androgen and thyroid systems, the so-called endocrine disruptors (US Environmental Protection Agency, US-EPA: https://www.epa.gov/endocrine-disruption). The estrogen MOA allows for a transcriptional profiling approach to assess the potential estrogenic activity of chemicals of interest, when coupled to a robust and biologically relevant system capable of mounting a physiologically significant estrogenic response. We have used microarray technology to assess the transcriptional changes associated with estrogenic chemicals exposure and demonstrated the reliability of this approach in an in vivo model, the female reproductive tract of rats during embryo/fetal development and in the juvenile rat [35,36]. The juvenile rat is capable of mounting an uterotrophic response to estrogen exposure, a reversible modification of the morphology and physiology of the uterus used as a standard assay to test estrogenicity of different compounds in vivo [37,38]. As a proof of principle, we initially evaluated the time- and dose-dependent genomic response of the indicated tissues, after exposure to a chemical with weak estrogenic activity, bisphenol A (BPA), a natural phytoestrogen, genistein (Gen) with mid-estrogenic activity or to a potent pharmaceutical estrogen (as compared to the endogenous hormone estradiol), 17 α-ethynyl estradiol (EE). The analysis of the genomic response elicited by these three model chemicals showed robust similarities between the transcript profile induced by EE, BPA, and Gen and has allowed us to identify a set of genes responsive to estrogen exposure in vivo (molecular fingerprint) [39].

This uterine transcriptional response to estrogen exposure determined in vivo encompasses the gene expression changes induced in the different cell types that constitute this organ. Thus it is realistic to believe that using transcript profiling the genomic response of cultured individual cell types derived from

TABLE 6.1 Prototypical Estrogenic Chemicals Evaluated in the Ishikawa Cells, After 8, 24, and 48 h of Exposure

Chemical	Final Concentration (M)			
	vLow	Low	Mid	High
17 α-Ethynyl estradiol	10×10^{-12}	10×10^{-10}	10×10^{-8}	10×10^{-6}
Bisphenol A	10×10^{-11}	10×10^{-9}	10×10^{-7}	10×10^{-5}
Genistein	10×10^{-10}	10×10^{-8}	10×10^{-6}	10×10^{-4}

the uterus to estrogen exposure can be also identified, and should represent part of the response of this organ. Further, moving forward, the goal is to establish an in vitro system to determine the potential estrogenic activity of chemicals of interest, develop reliable HTS assays, and minimize animal testing studies for this MOA. Pursuing these goals, we have directed our efforts on an estrogen-responsive human-derived endometrial adenocarcinoma cell line, Ishikawa cells, and evaluated the time and dose-dependent genomic response of these cells, after exposure to the prototypical estrogenic chemicals evaluated in vivo BPA, Gen, and EE (Table 6.1; [40–42]).

The analysis of the genomic response elicited by these three model chemicals showed robust similarities between the transcript profile induced by EE, BPA, and Gen (Fig. 6.1) and has allowed us to identify a set of genes responsive to estrogen exposure (molecular fingerprint) in an in vitro system. Global analyses aimed at detecting genes consistently modified by the three estrogenic chemicals in the Ishikawa cells, identified 296 unique genes whose expression changed in the same direction across the three chemicals, although the magnitude of the change of some genes is different in the cells exposed to these three ER agonists. An example of the expression of one gene upregulated (transforming growth factor alpha, TGFA1) and one downregulated (SRY (sex determining region Y)-box 4 (SOX4) by EE, BPA, and Gen, is shown in Fig. 6.2. Both TGFA1 and SOX4 are genes, among others we have identified, whose expression is regulated by estrogen in vivo [43,44].

We have also compared the transcriptional response of our in vitro system to the three estrogenic chemicals with the genomic response induced in vivo by these chemicals, and found a strong correlation between them. For example, the expression of early growth response 1 (EGR1), hydroxysteroid 11-beta dehydrogenase 2 (HSD11B2), and Kruppel-like factor 4 (KLF4) were similarly upregulated by EE, BPA, and Gen in both the developing rat uterus/ovaries and the Ishikawa cells, whereas the expression of N-myc downregulated gene family (NDRG) member 2 (NDRG2), transgelin (TAGLN), and insulin-like growth factor binding protein 3 (IGFBP3) was downregulated

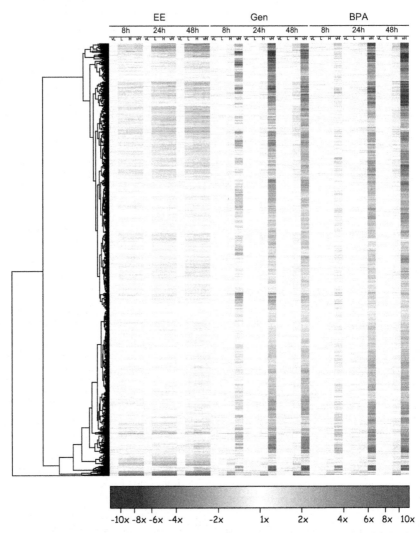

FIGURE 6.1   **Eisen diagram (or heat map) and a representative set of genes whose expression is modified in a dose-dependent manner by exposure to 17 α-ethynyl estradiol (EE), genistein (Gen), and bisphenol A (BPA) in Ishikawa cells.** The effect of EE, BPA, and Gen, at equipotent doses, orders of magnitude apart, on the gene expression profile of the Ishikawa cells was evaluated after 8, 24, and 48 h of exposure, using the Affymetrix Human Genome U133 Plus 2.0 arrays. The expression of 989 unique genes (displayed) was affected by EE, Gen, and BPA in a statistical significant manner (t test, P ≤ .0001 and fold change ≤ ±1.5 in at least one time and/or dose point). The fold change is the average fold change on the expression of each gene compared to vehicle treated control (n = 5, in all dose = groups, for each chemical). Each cell of the data table is represented as a color-coded rectangle in which the color indicates the expression value of unaffected (white), upregulated (red) or downregulated (blue) genes (identity by the Gen Bank accession number or GO symbol, not shown).

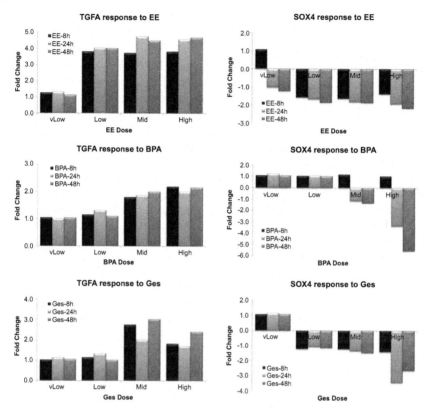

FIGURE 6.2 **Example of the robustness of the Ishikawa cells' response to chemical exposure, the response of two genes (probe sets), transforming growth factor alpha (TGFA1), and SRY (sex determining region Y)-box 4 (SOX4).** The expression of TGFA1 is upregulated by 17 α-ethynyl estradiol, genistein, and bisphenol A (shown as the mean fold change [n = 5] determined by comparing treated samples vs. the appropriate control, for all the dose (Table 6.1) and time points evaluated for each chemical). The expression of SOX4 (shown as the mean fold change [n = 5] determined by comparing treated samples vs. the appropriate control, for all the dose and time points evaluated for each chemical) follows the same pattern than TGFA1, but in opposite direction, because it is downregulated by chemicals with estrogenic activity.

in both the in vivo and the in vitro systems. These results have shown that the genomic response of a human-derived endometrial adenocarcinoma cell line (Ishikawa cells) to ER agonists exposure (EE, BPA, and Gen), mimics the in vivo response. This response is monotonic, time and dose dependent, as occurs in vivo, and has a clear overlap (molecular finger print) with the response elicited by the same chemicals (EE, BPA, and Gen, at equipotent concentrations) in vivo. These results indicate that the in vitro system we have selected, Ishikawa cells, is an appropriate model to determine estrogenic activity of chemicals of interest.

Although gene expression changes in response to estrogens is tissue, life stage, and sex specific, it is feasible to identify transcript profiles in an in vitro system that are diagnostic of this MOA. We hypothesized that chemicals with varied estrogenic activity would also change the expression of some or all of the genes affected by our model chemicals, but that nonestrogens would elicit changes in few or none of these genes. This in vitro approach in combination with the appropriate bioinformatics tools for ontological and pathway analyses to investigate the underlying molecular mechanisms of chemical action allows the screening of the estrogenic potential of chemicals of interest in a relatively fast manner and can be used to elucidate the shape of the dose–response curve at exposure levels below the no observed adverse effect level, an important need in risk assessment. Potential applications of this approach also include prioritizing chemicals for testing and designing customized testing approaches based on the chemical. These mechanism-specific transcript profiles can be used to define discrete pathways being affected by chemical exposure, and at the same time can be used as the basis for predictive toxicity (based on MOA), especially when grouping chemicals for read across purposes. This last application could be achieved by the use of the connectivity mapping method, originally described by Ref. [45].

Connectivity Mapping to Define Mode of Action

From multiple in vivo studies performed to address chemical toxicity, it has been observed that many chemicals have a number of biological effects, and it can be hypothesized that these effects are associated with specific cellular pathways. These data indicate that these chemicals elicit their biological actions through more than one MOA. For example, genistein has multiple biological activities (reviewed in Ref. [46]), including the inhibition of protein tyrosine kinases; topoisomerase II, and also inhibits enzymes involved in phosphatidylinositol metabolism. Genistein also regulates the signal transduction components Akt, FAK (or PTK2, protein tyrosine kinase 2), ErbB-2, and Bcl-2 [47]. Because of its structural similarity, genistein has been shown to compete with 17 β-estradiol for ER binding [28], resulting in agonistic or antagonistic activity, depending on the tissue being evaluated and specific interactions between the ERs and the proteins, coactivators, and transcription factors, that associate with them [48]. Genistein also arrests the cell cycle, induces apoptosis, has antioxidant activity, antiangiogenic, and antimetastatic properties, and has been shown to inhibit carcinogenesis in animal models [47,49]. Various in vivo and in vitro studies have shown that genistein can regulate the expression of multiple genes, whose products are involved in the regulation of multiple biological processes, including cell growth, cell cycle, cell signal transduction, angiogenesis, tumor cell invasion, and metastasis [35,42,50,51]. Therefore, exposure to genistein results in changes in multiple systems and acts not just by modifying the estrogen system, but various pathways, actions that can be identifiable by the transcriptional changes it elicits [35,42].

In order to make maximal use of gene expression data, it is important to have a database representing a broad spectrum of gene expression profiles elicited by chemicals with specific MOA that can be queried using the transcription data from a new compound. The ultimate goal of this approach is not to determine every mechanistic detail of toxicity associated with the exposure to a particular chemical, but to obtain sufficient information about the biological activity of the chemical that allow the identification of its potential toxic action. This biological activity is represented within a characteristic gene expression profile, as we have shown with chemicals with estrogenic activity in Ishikawa cells. However, the same transcriptional profile can be used to determine whether a given chemical acts throughout multiple MOAs. We have proposed the use of connectivity mapping (cMAP) to explore chemical MOA [5]. This approach (cMAP) can be used to identify connections between biological activities associated with specific chemicals and if they have been associated with well-defined MOAs, guide the identification of MOAs for chemicals of interest. It has to be stressed that the critical component needed to be able to apply this approach is to have access to a robust database of transcriptional signatures associated with a broad set of chemicals, covering all the potential MOAs, which will allow a better understanding of the molecular pathways relevant to assess their potential toxicity for humans and environmentally relevant species. We and others are working to build such a database ([5]; the Broad Institute of MIT and Harvard (http://www.broad.mit.edu/cmap/)).

In vitro to in Vivo Extrapolation

One of the limitations to using in vitro data in a risk assessment context is that determining the in vivo dosage that is equivalent to an active in vitro concentration is not straightforward. There has been a considerable amount of recent research on in vitro to in vivo extrapolation in order to address this problem.

Some of the critical data require chemical-specific parameters to predict in vivo absorption, distribution, metabolism, and excretion (ADME) in order to be able to build physiologically based toxicokinetic models [52]. Prediction of the biological activity of any given chemical, in vivo, requires the integration of data on the MOAs with data on pharmacokinetic behavior of such chemical. Models have been developed that combine in silico predictions of chemical behavior based on physical chemistry with laboratory measurements of hepatic clearance (including metabolism) and partition coefficients between external and biological matrices (e.g., air:blood, blood:tissue). For metabolism assessment the options could be to use microsomal fractions (from liver, plasma, lung or skin), primary hepatocytes, pluripotent stem cells differentiated into hepatocytes, or even liver slices [53,54]. If the main route of exposure is dermal, then the metabolism should be assessed also in human skin models, such as native skin samples, or 2D and 3D skin cultures [55–57]. To assess chemical elimination (detoxification) pathways, liver, kidney, intestinal, skin, and lung models can be explored [53,54,57,58].

The most valuable type of data that transcriptional profiling can contribute to the in vitro to in vivo extrapolation process is the identification of molecular pathways associated with the cellular response to chemical exposure and the concentrations at which they occur [59]. Ideally, these data can be compared to the same metric (i.e., altered expression of specific genes) in vivo that is associated with an adverse response, for one or more prototype compounds that act via the same MOA.

The great majority of in vitro systems currently available are static. When a chemical is being tested in such a system, this chemical is added to the culture medium and, apart from processes like evaporation, binding to plastic cell culture material or the proteins in the media, or a possible metabolism in the cell culture, remains there until the next medium change [60]. This is quite different from the situation in vivo where the blood flow, excretion processes and metabolism result in a more dynamic situation. The development of microfluidics systems and organs-on-a-chip models are promising tools to address the shortcomings of the static in vitro systems and have the potential to produce platforms to better assess the biological activity of any given chemical [61], particularly when these microphysiological systems (chips) incorporate the representation of multiple organs in one single unit and each "organ" is able to display its main functions. However, this field is still at an early stage of development and major challenges need to be addressed prior to the embracement of these technologies by the toxicology field. The ultimate goal in organs-on-a-chip research for chemical toxicity assessment is to develop reliable systems, whereby several organ cultures, representing all the main human organs, can mimic complex interactions between the organs and body systems in a dynamic microphysiological environment and provide in vivo-like pharmacokinetics and pharmacodynamics parameters for chemical ADME assessment.

CONCLUSION

In this chapter, an approach that focuses on the use of transcriptional profiling in an in vitro system to assess the estrogenic potential of chemicals of interest has been presented. The existing data indicate that the Ishikawa cells are capable of mounting a proper estrogenic response, as compared with the in vivo response, as it has been proven with estrogenic chemicals of varied potency. These cells offer a robust in vitro system to evaluate the estrogenic activity potential of chemicals of interest and at the very least can be used as a preliminary screening for this MOA or to prioritize chemicals for further testing, or as a weight of evidence for decision making. This in vitro system is amenable to dose– and time–response evaluations and allows the assessment of chemical exposure at concentrations that are occupational, consumer, and/or environmental relevant, critical parameters to generate usable data for the risk assessment process.

Addressing chemical safety in an in vitro system has advantages and disadvantages and results can be obtained at a very fast pace, even in a HTS format; if human-derived cells are used then there is no need to extrapolate data. However, these simple systems (one single cell type) lack the physiological complexity of an organ, or even a tissue, and thus have limited absorption, distribution, metabolism, and excretion capabilities [62], and the inclusion of transcriptional profiling after chemical exposure in the assays does not overcome these limitations. However, when the in vitro systems are used in a targeted fashion and implemented into integrated testing strategies to address specific questions (designed with a "fit for purpose" objective), a toxicogenomic approach, used in combination with bioinformatics tools for ontological and pathway analyses to investigate the underlying molecular mechanisms of chemical action, provides detailed biological information that can be used to define key events along a biological pathway affected by chemical exposure, which potentially could end with an adverse outcome. Because chemicals that elicit similar biological effects also modify the expression of similar genes, or biological pathways, a transcriptional profiling approach can be used to gather information to define chemical groupings and validate read across assessments, providing the appropriate data to create mechanism or MOA-based structural activity relationships.

Transcriptional profiling after chemical exposure of an in vitro system is a robust approach to gather chemical-specific biological activity data and offers a promising solution to address chemical safety that is reliable, reproducible, ethical, cost-effective, and time efficient.

REFERENCES

[1] Berggren E, Amcoff P, Benigni R, Blackburn K, Carney E, Blaauboer BJ. The long and winding road of progress in the use of in vitro data for risk assessment purposes: from "carnation test" to integrated testing strategies. Toxicology 2015;5(332):4–7.

[2] Daston G, Knight DJ, Schwarz M, Gocht T, Thomas RS, Mahony C, et al. SEURAT: safety evaluation ultimately replacing animal testing – recommendations for future research in the field of predictive toxicology. Arch Toxicol 2015;89(1):15–23.

[3] Organisation for Economic Co-operation and Development. Guidance document on developing and assessing adverse outcome pathways. Series on testing and assessment, no. 184. Paris, France: OECD Environment, Health and Safety Publications; 2013.

[4] Ankley GT, Bennett RS, Erickson RJ, Hoff DJ, Hornung MW, Johnson RD, Mount DR, et al. Adverse outcome pathways: a conceptual framework to support ecotoxicology research and risk assessment. Environ Toxicol Chem 2010;29(3):730–41.

[5] De Abrew KN, Kainkaryam RM, Shan YK, Overmann GJ, Settivari RS, Wang X, et al. Grouping 34 chemicals based on mode of action using connectivity mapping. Toxicol Sci 2016;151(2):447–61.

[6] Deferme L, Wolters J, Claessen S, Briedé J, Kleinjans J. Oxidative stress mechanisms do not discriminate between genotoxic and nongenotoxic liver carcinogens. Chem Res Toxicol 2015;28(8):1636–46.

[7] Dionisio KL, Frame AM, Goldsmith MR, Wambaugh JF, Liddell A, Cathey T, et al. Exploring consumer exposure pathways and patterns of use for chemicals in the environment. Toxicol Rep 2015;2:228–37.

[8] Egeghy PP, Judson R, Gangwal S, Mosher S, Smith D, Vail J, et al. The exposure data landscape for manufactured chemicals. Sci Total Environ 2012;414:159–66.

[9] Muir DC, Howard PH. Are there other persistent organic pollutants? A challenge for environmental chemists. Environ Sci Technol 2006;40(23):7157–66.

[10] Nelson LR, Bulun SE. Estrogen production and action. J Am Acad Dermatol 2001;45(3 Suppl.):S116–24.

[11] Simpson ER. Sources of estrogen and their importance. J Steroid Biochem Mol Biol 2003;86(3–5):225–30.

[12] Edwards DP. Regulation of signal transduction pathways by estrogen and progesterone. Annu Rev Physiol 2005;67:335–76.

[13] Schulster M, Bernie AM, Ramasamy R. The role of estradiol in male reproductive function. Asian J Androl 2016;18(3):435–40.

[14] Khan D, Ansar Ahmed S. The immune system is a natural target for estrogen action: opposing effects of estrogen in two prototypical autoimmune diseases. Front Immunol 2016;(6):635. http://dx.doi.org/10.3389/fimmu.2015.00635.

[15] Menazza S, Murphy E. The expanding complexity of estrogen receptor signaling in the cardiovascular system. Circ Res 2016;118(6):994–1007.

[16] Rossetti MF, Cambiasso MJ, Holschbach MA, Cabrera R. Estrogens and progestagens: synthesis and action in the brain. J Neuroendocrinol 2016;28(7). http://dx.doi.org/10.1111/jne.12402.

[17] Hall JM, Couse JF, Korach KS. The multifaceted mechanisms of estradiol and estrogen receptor signaling. J Biol Chem 2001;276(40):36869–72.

[18] Nilsson S, Mäkelä S, Treuter E, Tujague M, Thomsen J, Andersson G, et al. Mechanisms of estrogen action. Physiol Rev 2001;81(4):1535–65.

[19] Diel P. Tissue-specific estrogenic response and molecular mechanisms. Toxicol Lett 2002;127(1–3):217–24.

[20] Kushner PJ, Agard DA, Greene GL, Scanlan TS, Shiau AK, Uht RM, et al. Estrogen receptor pathways to AP-1. J Steroid Biochem Mol Biol 2000;74(5):311–7.

[21] Bodine PV, Harris HA, Komm BS. Suppression of ligand-dependent estrogen receptor activity by bone-resorbing cytokines in human osteoblasts. Endocrinology 1999;140(6):2439–51.

[22] Evans MJ, Harris HA, Miller CP, Karathanasis SK, Adelman SJ. Estrogen receptors alpha and beta have similar activities in multiple endothelial cell pathways. Endocrinology 2002;143(10):3785–95.

[23] Banerjee S, Chambliss KL, Mineo C, Shaul PW. Recent insights into non-nuclear actions of estrogen receptor alpha. Steroids 2014;81:64–9.

[24] Barton M. Not lost in translation: emerging clinical importance of the G protein-coupled estrogen receptor GPER. Steroids 2016;111:37–45.

[25] Prossnitz ER, Hathaway HJ. What have we learned about GPER function in physiology and disease from knockout mice? J Steroid Biochem Mol Biol 2015;153:114–26.

[26] Bengtson CP, Bading H. Nuclear calcium signaling. Adv Exp Med Biol 2012;970:377–405.

[27] Coleman KM, Dutertre M, El-Gharbawy A, Rowan BG, Weigel NL, Smith CL. Mechanistic differences in the activation of estrogen receptor-alpha (ER alpha)- and ER beta-dependent gene expression by cAMP signaling pathway(s). J Biol Chem 2003;278(15):12834–45.

[28] Kuiper GG, Carlsson B, Grandien K, Enmark E, Häggblad J, Nilsson S, et al. Comparison of the ligand binding specificity and transcript tissue distribution of estrogen receptors alpha and beta. Endocrinology 1997;138(3):863–70.

[29] Dix DJ, Houck KA, Martin MT, Richard AM, Setzer RW, Kavlock RJ. The ToxCast program for prioritizing toxicity testing of environmental chemicals. Toxicol Sci 2007;95(1):5–12.

[30] Judson RS, Magpantay FM, Chickarmane V, Haskell C, Tania N, Taylor J, et al. Integrated model of chemical perturbations of a biological pathway using 18 in vitro high-throughput screening assays for the estrogen receptor. Toxicol Sci 2015;148(1):137–54.

[31] Browne P, Judson RS, Casey WM, Kleinstreuer NC, Thomas RS. Screening chemicals for estrogen receptor bioactivity using a computational model. Environ Sci Technol 2015;49(14):8804–14.

[32] Huang R, Sakamuru S, Martin MT, Reif DM, Judson RS, Houck KA, et al. Profiling of the Tox21 10K compound library for agonists and antagonists of the estrogen receptor alpha signaling pathway. Sci Rep 2014;4:5664. http://dx.doi.org/10.1038/srep05664.

[33] Daston GP, Naciff JM. Predicting developmental toxicity through toxicogenomics. Birth Defects Res C Embryo Today 2010;90(2):110–7.

[34] Ozsolak F, Milos PM. RNA sequencing: advances, challenges and opportunities. Nat Rev Genet 2011;12(2):87–98.

[35] Naciff JM, Jump ML, Torontali SM, Carr GJ, Tiesman JP, Overmann GJ, et al. Gene expression profile induced by 17alpha-ethynyl estradiol, bisphenol A, and genistein in the developing female reproductive system of the rat. Toxicol Sci 2002;68(1):184–99.

[36] Naciff JM, Overmann GJ, Torontali SM, Carr GJ, Tiesman JP, Richardson BD, et al. Gene expression profile induced by 17 alpha-ethynyl estradiol in the prepubertal female reproductive system of the rat. Toxicol Sci 2003;72(2):314–30.

[37] OECD. Third meeting of the validation management group for the screening and testing of endocrine disrupters (mammalian effects). In: Joint meeting of the chemicals committee and the working party on chemicals, pesticides and biotechnology. Organization of Economic Cooperation and Development; 2001. http://www.oecd.org.

[38] U.S. EPA. Endocrine disruptor screening and testing advisory committee (EDSTAC). Final report. Screening and testingU.S. Environmental Protection Agency; 1998. [Chapter 5].

[39] Naciff JM, Daston GP. Toxicogenomic approach to endocrine disrupters: identification of a transcript profile characteristic of chemicals with estrogenic activity. Toxicol Pathol 2004;32(Suppl. 2):59–70.

[40] Naciff JM, Khambatta ZS, Thomason RG, Carr GJ, Tiesman JP, Singleton DW, et al. The genomic response of a human uterine endometrial adenocarcinoma cell line to 17alpha-ethynyl estradiol. Toxicol Sci 2009;107(1):40–55.

[41] Naciff JM, Khambatta ZS, Reichling TD, Carr GJ, Tiesman JP, Singleton DW, et al. The genomic response of Ishikawa cells to bisphenol A exposure is dose- and time-dependent. Toxicology 2010;270(2–3):137–49.

[42] Naciff JM, Khambatta ZS, Carr GJ, Tiesman JP, Singleton DW, Khan SA, et al. Dose- and time-dependent transcriptional response of Ishikawa cells exposed to genistein. Toxicol Sci 2016;151(1):71–87.

[43] Hunt SM, Clarke CL. Expression and hormonal regulation of the Sox4 gene in mouse female reproductive tissues. Biol Reprod 1999;61(2):476–81.

[44] Maekawa T, Sakuma A, Taniuchi S, Ogo Y, Iguchi T, Takeuchi S, et al. Transforming growth factor-α mRNA expression and its possible roles in mouse endometrial stromal cells. Zool Sci 2012;29(6):377–83.

[45] Lamb J, Crawford ED, Peck D, Modell JW, Blat IC, Wrobel MJ, et al. The connectivity map: using gene-expression signatures to connect small molecules, genes, and disease. Science 1929-35;313(5795).

[46] Polkowski K, Mazurek AP. Biological properties of genistein. A review of in vitro and in vivo data. Acta Pol Pharm 2000;57(2):135–55.

[47] Banerjee S, Li Y, Wang Z, Sarkar FH. Multi-targeted therapy of cancer by genistein. Cancer Lett 2008;269(2):226–42.

[48] Leitman DC, Paruthiyil S, Vivar OI, Saunier EF, Herber CB, Cohen I, et al. Regulation of specific target genes and biological responses by estrogen receptor subtype agonists. Curr Opin Pharmacol 2010;10(6):629–36.

[49] Sahin K, Tuzcu M, Sahin N, Akdemir F, Ozercan I, Bayraktar S, et al. Inhibitory effects of combination of lycopene and genistein on 7,12- dimethyl benz(a)anthracene-induced breast cancer in rats. Nutr Cancer 2011;63(8):1279–86.

[50] Li Y, Che M, Bhagat S, Ellis KL, Kucuk O, Doerge DR, et al. Regulation of gene expression and inhibition of experimental prostate cancer bone metastasis by dietary genistein. Neoplasia 2004;6(4):354–63.

[51] Di X, Andrews DM, Tucker CJ, Yu L, Moore AB, Zheng X, et al. A high concentration of genistein down-regulates activin A, Smad3 and other TGF-β pathway genes in human uterine leiomyoma cells. Exp Mol Med 2012;44(4):281–92.

[52] Yoon M, Campbell JL, Andersen ME, Clewell HJ. Quantitative in vitro to in vivo extrapolation of cell-based toxicity assay results. Crit Rev Toxicol 2012;42:633–52.

[53] Schepers A, Li C, Chhabra A, Seney BT, Bhatia S. Engineering a perfusable 3D human liver platform from iPS cells. Lab Chip 2016;16(14):2644–53.

[54] Godoy P, Hewitt NJ, Albrecht U, Andersen ME, Ansari N, Bhattacharya S, et al. Recent advances in 2D and 3D in vitro systems using primary hepatocytes, alternative hepatocyte sources and non-parenchymal liver cells and their use in investigating mechanisms of hepatotoxicity, cell signaling and ADME. Arch Toxicol 2013;87(8):1315–530.

[55] Danilenko DM, Phillips GD, Diaz D. Vitro skin models and their predictability in defining normal and disease biology, pharmacology, and toxicity. Toxicol Pathol 2016;44(4):555–63.

[56] Abdallah MA, Pawar G, Harrad S. Evaluation of in vitro vs. in vivo methods for assessment of dermal absorption of organic flame retardants: a review. Environ Int 2015;74:13–22.

[57] Wiegand C, Hewitt NJ, Merk HF, Reisinger K. Dermal xenobiotic metabolism: a comparison between native human skin, four in vitro skin test systems and a liver system. Skin Pharmacol Physiol 2014;27(5):263–75.

[58] Jacquoilleot S, Sheffield D, Olayanju A, Sison-Young R, Kitteringham NR, Naisbitt DJ, et al. Glutathione metabolism in the HaCaT cell line as a model for the detoxification of the model sensitisers 2,4-dinitrohalobenzenes in human skin. Toxicol Lett 2015;237(1):11–20.

[59] Patlewicz G, Simon T, Goyak K, Phillips RD, Rowlands JC, Seidel SD, et al. Use and validation of HT/HC assays to support 21st century toxicity evaluations. Regul Toxicol Pharmacol 2013;65:259–68.

[60] Blaauboer BJ. The long and winding road of progress in the use of in vitro data for risk assessment purposes: from "carnation test" to integrated testing strategies. Toxicology 2015;332:4–7.

[61] Jackson EL, Lu H. Three-dimensional models for studying development and disease: moving on from organisms to organs-on-a-chip and organoids. Integr Biol (Camb) 2016;8(6):672–83.

[62] Versteeg DJ, Naciff JM. In response: ecotoxicogenomics addressing future needs: an industry perspective. Environ Toxicol Chem 2015;34(4):704–6.

Chapter 7

Extrapolation of In Vitro Results to Predict Human Toxicity

Sonali Das

Syngene International Pvt. Ltd., Bangalore, India

NEED TO UNDERSTAND TOXICITY

The hallmark for safety assessment by toxicologist follows 500 years old Paracelsus' principle [1]. "All things are poison and nothing is without poison. Solely the dose determines that a thing is not a poison."

Toxicity is not only caused by known toxins, it can very well occur as a side effect of an effective drug or a compound. Originating from a small perturbation in one of the million biological pathways that are part of a living system, toxicity can be spread to damage the whole system to a point of no return. Therefore, identifying first and then understanding the root of derangement will help to implement the on time damage control that can prevent the long-term irreversible impairment of the system. Several methods have been developed to predict toxicity in various ways and up to different extents.

In order to minimize the toxic side effect of a potential drug candidate various screening methods are followed by the pharmaceutical companies during the drug development phases. The key challenge in drug development as of today is accurate identification and determination of associated human toxicity [2–4]; major reasons for drug withdrawal during phase I–III testing.

The cost incurred to bring a successful drug in the market has gone up to as high as $800 million with an estimated time duration of 12–15 years [5]. After going through this exhaustive exercise involving huge expenditure and effort the number of drugs that are withdrawn from the market and number of drugs getting black box warning in the past three decades clearly indicates that there is a sincere need to develop an approach to predict accurate human toxicity to stop postmarketing withdrawal and not to market drugs with adverse toxicity. A recent study projected that the cost for making a successful drug candidate can be reduced by ~$350 million by increasing the clinical success rate from 1 in 5 to 1 in 3. An additional reduction of $129 million (25% of the total expenditure) can be introduced by a reduction in total development pipeline and review by

In Vitro Toxicology. http://dx.doi.org/10.1016/B978-0-12-804667-8.00007-9

the regulatory authority. Considering the huge monetary loss due to late stage failure of drug development (phase II and III), need for a full-proof predictive capability in identifying a safe NCE (new chemical entity)/ new biological entity become a very important present day need [2][5,6].

In this article, we will address caveats in the present screening pipeline that lie behind the failure of predicting toxicity in human at the very late stage of drug development. We will emphasize on the technologies involving methods to streamline animal testing.

ORGAN LEVEL AND SYSTEM LEVEL TOXICITY

Liver, kidney, and heart are the three major organs that are majorly monitored during toxicity evaluation in the drug development pipeline. During 1975–1999, 45 of the approved drugs have been marked with one or more black box warning and 16 were withdrawn from the market due to safety issue [5]. Twenty-two percent of the drugs having black box warning and 31% of withdrawn drugs were associated mainly with liver toxicity. Thus hepatotoxicity is becoming the major cause behind postmarketing withdrawal of already approved drugs. Table 7.1 enlists the drugs/chemicals that were withdrawn from the market due to associated toxicity observed in clinical trials during 1974–2015 [7–10].

The safety estimation by toxicologists mainly governs by dose–response relationship that typically ignores the mechanistic insight. Toxic effect of a compound would impair cellular homeostasis that in turn causes organ damage reflected in elevated levels of specific biomarkers present in biological fluids; level of these markers are often correlated with extent of damage at the system level. Released markers are typically an indicator of late stage damage of the organ of origin.

Owing to the easy noninvasive access of body fluids, plasma and urine biomarkers are routinely monitored during preclinical evaluation. A set of organ-specific toxicity biomarkers are listed below:

- Plasma marker for hepatotoxicity—ALT (alanine aminotransferase), AST (aspartate aminotransferase), GST (glutathione S-transferases), bilirubin, ALP (alkaline phosphatase), bile salts
- Plasma and urine marker for nephrotoxicity—BUN (blood urea nitrogen), creatinine, uCr, Kim-1 (kidney injury molecule-1), lipocalin 2, OPN, β2M, cystatin C, albumin, GGT (gamma-glutamyl transferase), GST, NAG (N-acetyl-D-glucosaminidase), NGAL (neutrophil gelatinase-associated lipocalin), cyclopholin, IL18 (interleukin-18)
- Plasma marker for cardiotoxicity—To date, only ischemia-modified albumin is approved by the FDA, using the albumin cobalt-binding test, for assessment of myocardial ischemia

HEPATOTOXICITY

Defects in various intracellular biological processes culminate in organ level toxicity in an integrated manner. Necrosis, steatosis, oxidative stress, and

TABLE 7.1 Summary of Drug Withdrawal During 1974–2015

Drug	Indication	Type of Toxicity	Year Withdrawn
Ximelagatran	Oral anticoagulant	Hepatotoxic	2006
Cylert	Attention deficit hyperactive disorder and necrolepsy	Hepatotoxic	2005
Nefazodone	Depression	Hepatotoxic	2003
Troglitazone	Anti-inflammatory and antidiabetic	Hepatotoxic	2000
Duract	Nonsteroidal anti-inflammatory drug (NSAID)	Hepatotoxic	1998
Ebrotidine	Gastroduodenal lesions	Hepatotoxic	1998
Tolcapone	Parkinson disease	Hepatotoxic	1998
Tolrestat	Diabetic nephropathy	Hepatotoxic	1996
Chlormezanone	Anxiety disorders	Hepatotoxic	1996
Alpidem	Anxiety disorders	Hepatotoxic	1994
Bendazac	NSAID	Hepatotoxic	1993
Dilevalol	Hypertension	Hepatotoxic	1990
Fipexide	Psychostimulation	Hepatotoxic	1990
Tiryanfen	Diuretic and uricosurin	Hepatotoxic	1982
Temafloxacin	Bacterial infection	Renal toxic	1992
Nomifensine	Antidepressant	Renal and hepatotoxic	1986
Methoxuflurane	Anesthetic	Renal and hepatotoxic	1974 (in use in Australia)
Sorivudin	Herpes simplex and vericella-zoster virus infection	Gastric toxicity	1993
Pergolide	Parkinson's neurodegeneration	Cardiac toxicity	2007

Continued

TABLE 7.1 Summary of Drug Withdrawal During 1974–2015—cont'd

Drug	Indication	Type of Toxicity	Year Withdrawn
Adderall-XR	Attention deficit hyperactive disorder	Cardiac toxicity	2005
Vioxx	Acute and chronic pain	Cardiac toxicity	2004
Levaacetylmethadol	Treatment of opioid addiction	Cardiac toxicity	2003
Droperidol	Premedication for anesthesia	Cardiac toxicity	2001
Cisapride	Gastrointestinal reflux	Cardiac toxicity	2000
Hismanal	Antihistamine	Cardiac toxicity	1999
Encainide	Antiarrhythmic	Cardiac toxicity	1999
Seldane	Antihistamine	Cardiac toxicity	1998
Terfenadine	Allergies	Cardiac toxicity	1997–99
Fenfluramine	Appetite suppressant	Cardiac toxicity	1997
Flosequinan	Vasodialator	Cardiac toxicity	1993
Terodiline	Urinary inconsistence	Cardiac toxicity	1992
Grepafloxacin	Bacterial infection	Cardiac toxicity	1997
Alosetron	Irritable bowel syndrome	Ischemic colitis	2000
Mibefradil	Calcium channel blocker	Drug–drug interaction	1998
Cerivastatin	Cholesterol lowering	Rhabdomyolysis	2001
Zomepirac	NSAID	Anaphylactic reaction, nonfetal allergic reactions, renal failure	1983
Fasiglifam (TAK-875)	Antidiabetic	Hepatotoxicity	2013

cholestasis are the four classical drug induced liver injury (DILI) observed as a clinical outcome. These are the resultants of altered lipid and phospholipid metabolism, reactive intermediates turnover, mitochondrial toxicity, impaired synthesis of energy (adenosine triphosphate [ATP]), and intracellular Ca-homeostasis [11]. Cellular level derangement results in the released biomarkers present in body fluids. Lack of mechanistic correlation between the biomarker and cellular level derangement often leads to an incomplete picture on the effect of the compound and how the system responses to change.

To bridge the gap between the outcomes obtained at the molecular level (from in vitro cell system) and from whole organ, many academic and industrial organizations are coming up with three-dimensional (3D) architecture mimicking organ behavior; industries offer a battery of assays that integrates in vitro and in vivo outcomes. These upcoming endeavors can help to map the mechanistic rationale for organ toxicity at the molecular level. With the establishment of a validated and then tested correlation between cell level observation and organ (3D model) predicted outcome, a modified strategy for animal testing methodology can be envisaged [12].

Table 7.2 [13–27] lists industries that are offering various services to be used by pharmaceuticals in the NCE screening pipeline (better and enhanced alternative to cell monolayer based liver toxicity assessment).

TOXICITY ASSESSMENT IN VITRO: ADVANTAGES AND DISADVANTAGES

In Vitro Systems for Liver Toxicity Assessment

Routinely used in vitro test systems by drug companies include HepG2, H4IIE, THLE, HepaRG cell lines (liver specific). Embryonic stem (ES) cells and induced pluripotent stem cells (iPSC) derived hepatocyte (for both rat and human) are also in use extensively as in vitro test system for toxicity screening [28,29]. Inamura et al. described an efficient method of hepatoblast formation from human ES and iPSC. Proliferation from hepatoblasts differentiates into both hepatocyte and cholangiocytes [30]. To mimic the three-dimensional structure of liver biology with embedded zonal diversity, researchers are coming up with various forms of high throughput chip.

Although research laboratories are using various cell lines and isolated mitochondria, freshly isolated primary hepatocyte are still considered as a gold standard for in vitro toxicity assessment. Primary hepatocyte from rat and human are used routinely to distinguish the species specific responses. Advantages of using cell lines and monolayer primary cells lie in their easy access and quick turnaround time with the expected results. Experiments done using isolated mitochondria lack the complexity of having membrane permeability barrier and drug metabolism machinery to effectively mimic in vivo drug exposure [31–33]. Primary cells in monolayer culture do not mimic the zonal behavior of liver. Long-term treatment gets limited due to differentiation of primary cells in tissue culture condition. Both phenotypic and genotypic characteristics of liver

TABLE 7.2 Technologies to predict DILI

Type of Biochemical Measurement	Oxidative Stress	Steatosis and Phospholipidosis	Xenobiotic Metabolism	Mitochondrial Behavior	Cell Death Measurement
Heptox—DILI prediction platform (http://www.syngeneintl.com/services/bioinformatics/the-heptox-platform)	GSH/GSSG, Y—GCS, GR, ROS	MTP, FAS, CD36, CPT1	UGT	Complex 1 and 2, Mitochondrial membrane potential	ATP
Ceetox (Cyprotex acq) (http://www.cyprotex.com/)	ROS (DCFDA), GSH	LXR (claimed as xenobiotic metabolism) (involved in lipid and glucose homeostasis), phospholipidosis, steatosis	GST (membrane integrity), UGT, cytochrome P450 inhibition assay, pregnane X receptor, constitutive androstane receptor (CAR)	Mitochondrial toxicity (Glu/Gal assay), functional mitochondrial toxicity assay (using Seahorse XFe96 flux analyzer), mitochondrial potential, mitochondrial mass	ATP, LDH (membrane integrity), ALP, cell loss, cell membrane permeability cytochrome C, caspase 3/7 phospho-p53
Hepregen (www.hepregen.com/)	GSH (they refer this to reactive metabolites)	Lipid phospholipid accumulation	CYP450	ATP, MTT	Urea, albumin, apoptosis caspases
Fluofarma (www.fluofarma.com/)	Redox state, intracellular GSH content, GSH. NADPH depletion, ROS	Lipid accumulation, LDL accumulation, NADH/FAD content NEUTRAL/POLAR	Cyp induction, nuclear receptor interactions, microsomal stability	Mitochondrial toxicity intracellular Ca^{2+} mitochondrial function	Cytolysis, autophagy, p53 caspase 3/7, cytotoxicity mitochondrial depolarisation
Hemogenix (www.hemogenix.com/)	GSH, ROS assay	Lipid peroxidation assay	Cytochrome p450	ATP bioluminiscence assay^{2+}	Caspase 3 and 3/7 apoptosis assays, MTT, ATP bioluminescence assay
Pfizer (www.pfizer.com/)	GSH content, GSH area, and GSH average pixel intensity	lipid intensity		Mitochondrial Membrane Potential	

Type of Biochemical Measurement	Cholestasis	Other Assays		3D Culture	Cell System Used
Enzo (www.enzolifesciences.com/)	ROS—ID Total ROS/Superoxide detection kit, ROS/RNS detection kit, NO detection kits, heme oxygenase ELISAs	LYSO—ID Red cytotoxicity kit (measures lysosome accumulation)	SCREEN-WELL toxicity libraries (cytochrome p450 inactivation, Toxic metabolites)	Mitochondrial toxicity kit-MITO-ID extracellular O_2 sensor kits	Cytochrome C, p53, Survivin and Bcl-2 ELISAs, NUCLEAR–ID Cell Viability reagents CYTO–ID Long–term tracer Kits p21 and p27 Kip1 ELISAs
Insphero/Pharmacelsus (http://insphero.com/)	GSH/GSSG	Nile—Red Staining Oil—Red—O Staining	Cytochrome p450 enzyme activity, CYP3A4	Mitochondrial toxicity testing and microtissues for hepatotoxicity testing–2	ATP, α-GST ELISA Albumin ELISA LDH assay—
Bioreliance (www.bioreliance.com/)		Steatosis assay (oil red O is used to stain neutral lipids in the metabolically			GreenScreen HC assay measures induction of p53α
Qualyst (www.qualyst.com/)			p450 induction, p450 + transporter induction	Mitochondrial membrane potential, ATP	Hepatotoxicity kits: viability (LDH)
Heptox–DILIprediction platform (http://www.syngeneintl.com/services/bioinformatics/the-heptox-platform)				No	HepG2, H4IIE, Rat Hepatocyte
Ceetox (Cyprotexacq) (http://www.cyprotex.com/)	MDR1—MDCK, P—glycoprotein, BCRP, OAT,OCT, MATE, BSEP			Yes	HepG2
Hepregen (www.hepregen.com/)	Transporter kit	Cytokine immune response, ALT, bile secretion		yes	96–well format hepatocyte co–culture for Human, rat, monkey and dog

Continued

TABLE 7.2 Technologies to predict DILI—cont'd

Type of Biochemical Measurement	Cholestasis	Other Assays	3D Culture	Cell System Used
Fluofarma (www.fluofarma.com/)	Bile caniliculi toxicity, hepatobiliary function, Cholestasis/bile canaliculi network toxicity, MRP2	Flow cytometry and automated—high content imaging, intracellular Ca^{2+}, heme oxygenase1 (Ho-1), quinone oxidoreductase-1 (Nqo1), multidrug-resistant proteins (MRPs)	No	Hepatic tissue, HepG2, hepatoma cell lines, primary cultures of rat and human hepatocytes—sandwich culture
Hemogenix (www.hemogenix.com/)		LDH for membrane integrity, caspase detection, cellular drug—drug interactions, oxidative DNA damage, mitochondrial toxicology	No	Human, nonhuman primate, dog, rat, mouse; hepatocytes, iPSC derived hepatocytes, HepG2
Pfizer (www.pfizer.com/)		Human hepatocyte imaging assay technology: nuclei count, nuclei area, TMRM intensity	No	Human hepatocytes
Enzo (www.enzolifesciences.com/)	SCREEN–WELL toxicity libraries (cholestatic effect)	Hepatotoxicity libraries that contain compounds with toxicity profiles for steatosis, cholestatic effects, mitochondrial toxicity, CYP450 inactivation, toxic metabolites, Mallory body formation, elevation of liver enzymes, nonhepatotoxic controls.	No	
Insphero/Pharmacelsus (http://insphero.com/)	CaCO2 cell monolayers, 3D culture for bile transporters, CLF, and CMFDA secretion	MRP2, BSEP transporter, Presence of albumin	Yes	Human (HepaRG) and Rat Hepatocytes,3D InSight, liver microtissues: human, rat, dog, monkey

Type of Biochemical Measurement	Cholestasis	Other Assays	3D Culture	Cell System Used
Bioreliance (www.bioreliance.com/)	Drug permeability across CaCO$_2$ cell monolayer		3D skin available. Can be extended to liver	
Qualyst (www.qualyst.com/)	Hepatobiliary disposition (hepatic uptake, biliary excretion, and biliary clearance). Cholestatic potential screening.	Route of Elimination, transporter based hepatic drug–drug interactions, and potential drug-induced hepatotoxicity. Transporters: NTCP, OATP, BSEP, P–GP, MRP2, and BCRP	No	HepatoPac, B–CLEAR (sandwich cultured hepatocytes: rat, dog, monkey, and human)
Solvo (www.solvobiotech.com/)	Efflux transporter assays, uptake transporter assays, (BCRP, MDR, OAT, OCT)	Transporters: NTCP, OATP, BSEP, pGP,MRP1, MRP2, MRP3 and MRP5, and BCRP.OAT1, OATP1B1, OATP1B3, and OCT2.	No	Transformed Cell line licensing–MDCKII, CHO, HEK293, Hepatopac

Y-GCS, gamma-glutamylcysteine synthetase; *ATP*, adenosine triphosphate; *BCRP*, transporter, breast cancer resistance protein; *BSEP*, bile salt export pump; *CMFDA*, 5-chloromethylfluorescein diacetate; *DCFDA*, 2,7′-dichlorofluorescin diacetate; *DILI*, drug induced liver injury; *GR*, glutathione reductase; *GSH*, reduced glutathione; *GSSG*, glutathione disulfide; *iPSC*, induced pluripotent stem cells; *LDH*, lactate dehydrogenase; *MATE*, multidrug and toxic compound extrusion; *MDCK*, Madin-Darby canine kidney cell line; *MDR1*, multidrug resistance protein 1, *MITO*, mitochondrial; *MRP1*, multidrug resistance protein 1; *MRP2*, multidrug resistance protein 2; *MRP3*, multidrug resistance protein 3; *MTP*, microsomal triglyceride transfer protein; *MTT*, 3-(4,5-dimethylthiazol-2-yl)-2,5-diphenyltetrazolium bromide; *NTCP*, sodium/taurocholate cotransporting polypeptide; *OAT*, organic anion transporter; *OATP*, anion-transporting polypeptide; *OCT*, organic cation transporters; *pGP*, p-glycoprotein; *ROS*, reactive oxygen species; *TMRM*, tetramethylrhodamine, methyl ester.

cells differ among three zones [34]. Enhanced level of bile salts and bilirubin in plasma is attributed to cholestasis. To predict cholestatic potential (rise in bile salts and bilirubin in plasma) various cell lines transfected with specific bile transporters are in use as test systems [35–37]. Extrapolating the in vitro effect on transporter to estimate alteration of bile salt and bilirubin concentration in plasma is nontrivial. Hence, screening compounds through these in vitro systems may end up filtering false negative compounds as safe for not causing cholestasis. Many a times, alteration in various parts of biology together culminates into toxic end points.

In Vitro System for Nephrotoxicity Assessment

A panel of markers integrating information on alteration in glomerular, tubular, and interstitial function is yet to come in practice for an early detection of nephrotoxicity [38]. Similar to identifying liver toxicity to detect kidney alteration, there are various in vitro systems that are used routinely to predict nephrotoxicity outside an animal model.

Proximal tubular and distal tubular primary kidney cells from rats and humans are used to study drug metabolism, membrane transport and biochemical alterations during evaluation of compound-induced renal toxicity [37,38]. Commercially available human cell types, e.g., HK2 (human kidney-2) and hRPTEC (primary human renal proximal tubule epithelial cells) are used to understand kidney damage as an effect of drug treatment. Huang et al. compared the effect of a set of nephrotoxic compounds on both immortalized and primary cells and established a good correlation between in vivo and in vitro findings [39,40]. Table 7.3 lists the comparative outcome of toxicity testing using in vitro and in vivo systems.

TO OVERCOME IN VITRO ASSAY LIMITATION

A set of assays are performed to measure the extent and type of cell cytotoxicity caused by the test compound. This set comes under first cut of safety measure.

- Loss of cell membrane integrity induced cytolysis—estimated by lactate dehydrogenase release and membrane impermeable DNA dye;
- Apoptosis—estimated by caspase activation;
- Depletion of energy currency of cell, ATP, and oxidative stress marker, reduced glutathione;
- Reduction in cellular metabolic activity by tetrazolium salt assay and Alamar blue assay;
- Reduced cell proliferation by estimating inhibition in DNA and protein synthesis.

These assays end up estimating cell health when the damage caused by the treated compound is irreversible. This set of assays does not enlighten on the

TABLE 7.3 Disadvantages Associated With In Vitro Toxicity Identification

In Vivo Toxicity	In Vitro Toxicity
Originate from one organ damage or multiorgan damage	Originate from one type of in vitro cell system
Altered immune response contributes to systems level toxicity	Confined to toxicity identification in one cell type without the presence of immune system
Drug and drug metabolite induced toxic effects are captured	Cell lines as well as many primary cells when cultured in vitro are compromised in expressing basal level of drug metabolizing enzymes. Hence, mostly the effect of parent drug is captured.
Represent the cumulative derangement in both known and unknown physiological events	Toxicity assessment typically follow a well identified and specific pathways of cellular damage
Pharmacokinetics (PK) and pharmacodynamics (PD) based processing of the parent drug, after it enters the system, determines the drug exposure seen by various organs	• In vitro toxicity assessment is mainly a function of treatment concentration. Available drug amount is determined by the stability of the drug in culture media and the serum protein (when present in culture media) binding affinity of the drug. • Culturing cells in vitro in absence of low protein or in serum free condition for drug testing may end up testing toxicity at a very high concentration that can never be achieved in in vivo [41–43].
In vivo tissue architecture determines cellular differentiation (due to exposure of a compound), gene expression pattern, cell polarization, morphology as well as functional capability ([44–47]).	Cells grown in monolayer culture are in an environment where cell to cell communication among various cell types is typically absent. In vitro systems cannot represent the multi cellular level connectivity.
Various extent of biotransformation as well as generation of reactive metabolites are linked to various types of cellular toxicity including idiosyncratic toxicity.	Identification of the root cause of idiosyncratic toxicity is completely missing in an in vitro screening system.

onset of damage neither the mechanistic insight but measured the terminal toxic effect. Understanding the progression of toxicity from basal condition is important to find out the root cause of prevention. A relation between the early detection (by in vitro assays) and late stage biomarker may be established to prevent withdrawal of drug during clinical trials. The following are a few suggested ways:

- Comprehensive analysis of the mechanism behind complex cell cytotoxicity would require many different assays to run in parallel. Only then, the built in redundancy in biological system will reveal the progress of insult from prelethal to apoptosis and then finally necrosis.
- Multiple cell types and cell lines need to be tested simultaneously originating from different organs, e.g., liver, kidney, heart, lung, to get a holistic understanding.
- Specific target enzymes behind organ toxicity need to be included in the assay panel to elucidate the system behavior.
- Use of cell-based assays as opposed to target enzyme assay in isolation will mimic the in vivo environment when exposed to a drug. Protein binding affinity of a drug molecule governs the available effective concentration of free drug in the system. Whole-cell treatment in culture media having serum would represent the free and bound drug combination as observed in whole animal.
- Disease and toxicity sensitivities can be dictated by individual genetic variation. Coordinated effort in combining multidisciplinary teams generating information from many sources and technologies will help to integrate gene to protein to activity-based phenotype.

EXTRAPOLATION OF IN VITRO INFORMATION TO IN VIVO FOR HUMAN: SUCCESSFUL AND UNSUCCESSFUL OUTCOMES

The discrepancy due to extrapolating animal outcome to human falls majorly in the following facts:

- Metabolism differs between the two species.
- Animal studies done inside the laboratory setup under controlled experimental environment are unable to mimic the enormous diversity present in vast human population.
- Person specific genetic makeup, environmental effect, existing disease environment are a few factors that are often responsible for adverse and unpredicted drug reaction.

An organ is not merely a mass of cells organized in a particular fashion. Liver and kidney are the organs where toxicity induced by an NCE first becomes apparent before the compound is tested in humans.

UPCOMING POTENTIAL METHODS TO PREDICT TOXICITY

Comparative studies have been done to find out the correlation between nature of toxicities observed in regulatory animals and humans. Although there is a fair chance of having similar responses from both the systems for cardiovascular, hematologic, and gastrointestinal toxicities, hypersensitivity, cutaneous reactions, and hepatotoxicity signals differ a lot between the two test systems [48].

The toxicology paradigm needs to be evolved from the traditional experimental research techniques to investigational scientific research mode. Toxicity prediction during preclinical studies, and, identification of toxicity induced alteration or injury in humans is needed to streamline the drug development process, make it more efficient, and enhance patient safety during clinical studies. Normal and disease models of animals are used to demonstrate target engagement before clinical trial. Such endeavors are to be regarded as exploratory research and not definitive. Appropriate considerations need to be implemented before extrapolating such information to humans.

Systems biology based in silico approaches to model human diseases is an emerging alternative to laboratory animal usage. This can help to overcome the ethical issues associated with live animal usage [49,50]. Computation heavy simulation based prediction tools to characterize chemicals and potential adverse effects are gaining increasing importance in present days. Two such endeavors are DILIsym and Heptox.DILIsym is an in silico-based simulation platform to predict DILI. It is a multiscale and "middle out" platforms catered to predict compound-induced liver toxicity by integrating in vitro data, PK/PD data, and population heterogeneity [51–53]. Many pharmaceutical organizations came together to form a consortium to evaluate the potential of DILIsym prediction platform; an effort originated from Hamner Institute in North Carolina [52–55]. Heptox is an integrative platform to use in vitro data as input to run in silico simulation and predict in vivo outcome of a drug having potential to cause liver toxicity [13]. This patented technology (US Patent No. 8,645,075) originates from a Bioinformatics industry, Strand Life Sciences, Bangalore, India. At the core of Heptox platform there is a Virtual liver that mimics the function of normal liver physiology in computer. The results of a battery of in vitro assays are used as input to the model to simulate the effect of test compound on liver. Heptox not only flags the compounds or drugs that are likely to have liver related problems, it can also address the exposure-based impact in liver, an ever challenging questions to be resolved during preclinical toxicity screening [13]. Availability of both human and rat liver metabolism captured in in silico model can help to differentiate species specificity during preclinical screening stage.

REFERENCES

[1] Borzelleca JF. Paracelsus: herald of modern toxicology. Toxicol Sci 2000;53:2–4.
[2] Dorato MA, Buckley LA. Toxicology testing in drug discovery and development. Current protocols in toxicologyJohn Wiley & Sons, Inc.; 2007.
[3] Li AP. Accurate prediction of human drug toxicity: a major challenge in drug development. Chem Biol Interact 2004;150:3–7.

[4] Giezen TJ, Mantel-Teeuwisse Ak, Straus SM, Schellekens H, Leufkens HG, Egberts AC. Safety-related regulatory actions for biological approved in the Unites States and European Union. JAMA 2008;300:1887–96.

[5] DiMasi JA, Hansen RW, Grabowski HG. The price of innovation: new estimate of drug development cost. J Health Econ 2003;22:151–85.

[6] Kola I, Landis J. Can the pharmaceutical industry reduce attrition rates? Nat Rev Drug Discon 2004;3:711–5.

[7] Astashkinaa A, Mannb B, Graingera DW. A critical evaluation of in vitro cell culture models for high-throughput drug screening and toxicity. Pharmacol Ther 2012;134(1):82–106.

[8] MacDonald JS, Robertson RT. Toxicity testing in the 21st century: a view from the pharmaceutical industry. Tox Sci 2009;110(1):40–6.

[9] Keisu M, Andersson TB. Drug-induced liver injury in humans: the case of ximelagatran. Handb Exp Pharmacol 2009;196:407–18.

[10] Li X, Zhong K, Guo Z, Zhong D, Chen X. Fasiglifam (TAK-875) inhibits hepatobiliary transporters: a possible factor contributing to fasiglifam-induced liver injury. Drug Metab Dispos 2015;43:1751–9.

[11] Xu JJ, Diaz D, O'Brien PJ. Application of cytotoxicity assays and pre-lethal mechanistic assays for assessment of human hepatotoxicity potential. Chem Biol Interact 2004;150:115–28.

[12] Onyi Irrechukwu, Sonali Das, Rajeev Kumar, Stacy Krzyzewski, Zurawski Jr VR., McGeehan JK., Kalyanasundaram Subramanian, Strand Life Sciences, Bangalore, India, Hepregen Corporation, Medford, USA. 55th annual meeting of the society of toxicology 2016, New Orleans, Louisiana, USA. Oxidative stress response – DILI prediction connecting organ level response to cellular mechanism; 2016.

[13] Subramanian K, Raghavan S, Bhat AR, Das S, Dikshit JB, Kumar R, Narasimha MK, Nalini R, Radhakrishnan R, Raghunathan S. A systems biology based integrative framework to enhance the predictivity of in vitro methods for drug-induced liver injury. Expert Opin Drug Saf 2008;7(6):647–62.

[14] Leahy DE. Integrating in vitro ADMET data through generic physiologically based pharmacokinetic models. Expert Opin Drug Metabol Toxicol 2004;2(4):619–28.

[15] Cartmell J, Krstajic D, Leahy DE. Competitive Workflow: novel software architecture for automating drug design. Curr Opin Drug Disc Dev 2007;10(3):347–52.

[16] Rimann M, Laternser S, Gvozdenovic A, Muff R, Fuchs B, Kelm JM, Graf-Hausner U. An in vitro osteosarcoma 3D microtissue model for drug development. J Biotechnol 2014;189(10):129–35.

[17] Drewitz M, Helbling M, Fried N, Brieri M, Moritz W, Lichtenberg J, Keln JM. Towards automated production and drug sensitivity testing using scaffold-free spherical tumor microtissues. Biotechnol J 2011;6(12):1488–96.

[18] Khetani SR, Bhatia SN. Microscale culture of human liver cells for drug development. Nat Biotechnol 2008;26(1):120–6.

[19] Ukairo O, Kanchagar C, Moore A, Shi J, Gaffney J, Aoyama S, Rose K, Krzyzewski S, McGeehan J, Andersen ME, Khetani SR, LeCluyse EL. Long-term stability of primary rat hepatocytes in micropatterned cocultures. J Biochem Mol Toxicol 2013;27(3):204–12.

[20] Khetani SR, Kanchagar C, Ukairo O, Krzyzewski S, Moore A, Shi J, Aoyama S, Aleo M, Will Y. Use of micropatterned cocultures to detect compounds that cause drug-induced liver injury in humans. Toxicol Sci 2013;132(1):107–17.

[21] Oedit A, Vulto P, Ramautar R, Lindenburg PW, Hankemeier T. Lab-on-a-Chip hyphenation with mass spectrometry: strategies for bioanalytical applications. Curr Opin Biotechnol 2014;31(C):79–85.

[22] Kidney-on-a-Chip technology for drug-induced nephrotoxicity screening. Trends Biotechnol February 2016;34(2):156–70.

[23] Gunness P, Mueller D, Shevchenko V, Heinzle E, Ingelman-Sundberg M, Noor F. 3D organotypic cultures of human HepaRG cells: a tool for in vitro toxicity studies. Toxicol Sci 2013;133(1):67–78.

[24] Michaut A, Guillou DL, Moreau C, Bucher S, McGill MR, Martinais S, Gicquel T, Morel I, Robin M, Jaeschke H, Fromenty B. A cellular model to study drug-induced liver injury in nonalcoholic fatty liver disease: application to acetaminophen. Toxicol Appl Pharmacol 2016;292:40–55.

[25] Sáfár Z, Vaskó B, Ritchie TK, Imre G, Mogyorósi K, Erdő F, Rajnai Z, Fekete Z, Szerémy P, Muka L, Zolnerciks JK, Herédi-Szabó K, Ragueneau-Majlessi I, Krajcsi P. Investigating ABCB1-mediated drug-drug interactions: considerations for in vitro and in vivo assay design. Curr Drug Metab 2016;17(5):430–55.

[26] Perry CH, Smith WR, St Claire III RL, Brouwer KR. Automated applications of sandwich-cultured hepatocytes in the evaluation of hepatic drug transport. J Biomol Screen 2011;16(4):427–35.

[27] Xu JJ, Henstock PV, Dunn MC, Smith AR, Chabot JR, de Graaf D. Cellular imaging predictions of clinical drug-induced liver injury. Toxicol Sci 2008;105(1):97–105.

[28] Yu J, Vodyanik MA, Smuga-Otto K, Antosiewicz-Bourget J, Frane JL, Tian S, Nie J, Jonsdottir GA, Ruotti V, Stewart R, Slukvin II, Thomson JA. Induced pluripotent stem cell lines derived from human somatic cells. Science 2007;318:1917–20.

[29] Chen Y, Tseng C, Wang H, Kuo H, Yang VW, Lee OK. Rapid generation of mature hepatocyte-like cells from human induced pluripotent stem cells by an efficient three-step protocol. Hepatology 2012;55:1193–203.

[30] Inamura M, Kawabata K, Takayama K, Tashiro K, Sakurai F, Katayama K, Toyoda M, Akutsu H, Miyagawa Y, Okita H, Kiyokawa N, Umezawa A, Hayakawa T, Furue MK, Mizuguchi H. Efficient generation of hepatoblasts from human ES cells and iPS cells by transient overexpression of homeobox gene HEX. Mol Ther 2011;19(2):400–7.

[31] Krlhenbiihl S, Talosb C, Hoppeld UCL. Development and evaluation of a spectrophotometric assay for complex III in isolated mitochondria, tissues and fibroblasts from rats and humans. Clin Chim Acta 1994;230:177–87.

[32] Jones AJY, Hirst J. A spectrophotometric coupled enzyme assay to measure the activity of succinate dehydrogenase. Anal Biochem 2013;442:19–23.

[33] Fujikawa M, Yoshida M. A sensitive, simple assay of mitochondrial ATP synthesis of cultured mammalian cells suitable for high-throughput analysis. Biochem Biophys Res Com 2010;401:538–43.

[34] Jungermann K. Zonation of metabolism and gene expression in liver. Histochem Cell Biol 1995;103:81–91.

[35] Wang EJ, Casciano CN, Clement RP, Johnson WW. Fluorescent substrates of sister-P-glycoprotein (BSEP) evaluated as markers of active transport and inhibition: evidence for contingent unequal binding sites. Pharm Res April 2003;20(4):537–44.

[36] Morgan RE, Trauner M, van Staden CJ, Lee PH, Ramachandran B, Eschenberg M, Afshari CA, Qualls Jr CW, Lightfoot-Dunn R, Hamadeh HK. Interference with bile salt export pump function is a susceptibility factor for human liver injury in drug development. Toxicol Sci 2010;118(2):485–500.

[37] Murray JW, Thosani AJ, Wang P, Wolkoff AW. Heterogeneous accumulation of fluorescent bile acids in primary rat hepatocytes does not correlate with their homogenous expression of NTCP. Am J Physiol Gastrointest Liver Physiol 2011;301:G60–8.

[38] Fuchs TC, Hewitt P. Biomarkers for drug-induced renal damage and nephrotoxicity – an overview for applied toxicology. AAPS J 2011;13(4):615–31.

[39] Lash L. In vitro methods of assessing renal damage. Toxicol Pathol 1998;26(1):33–42.

[40] Lasha LH, Putta DA, Cai H. Drug metabolism enzyme expression and activity in primary cultures of human proximal tubular cells. Toxicology 2008;244(1):56–65.

[41] Fisher GJ, Duell EA, Nickoloff BJ, Annesley TM, Kowalke JK, Ellis CN, Voorhees JJ. Levels of cyclosporin in epidermis of treated psoriasis patients differentially inhibit growth of keratinocytes cultured in serum free versus serum containing media. J Invest Dermato 1988;91(2):142–6.

[42] Nagel SC, Vom Saal FS, Welshons WV. The effective free fraction of estradiol and xenoestrogens in human serum measured by whole cell uptake assays: physiology of delivery modifies estrogenic activity. Proc Soc Exp Biol Med 1998;217:300–9.

[43] Jones CF, Grainger DW. In vitro assessments of nanomaterial toxicity. Adv Drug Deliv Rev 2009;61:438–56.

[44] Hartmann F, Bissell DM. Metabolism of heme and bilirubin in rat and human small intestinal mucosa. J Clin Invest 1982;70(1):23–9.

[45] Cukierman E, Pankov R, Stevens DR, Yamada KM. Taking cell-matrix adhesions to the third dimension. Science 2001;294(5547):1708–12.

[46] Bissell MJ, Radisky DC, Rizki A, Weaver VM, Petersen OW. The organizing principle: microenvironmental influences in the normal and malignant breast. Differentiation 2002;70(9–10):537–46.

[47] Ghosh S, Spagnoli GC, Martin I, Ploegert S, Demougin P, Heberer M, Reschner A. Three-dimensional culture of melanoma cells profoundly affects gene expression profile: a high density oligonucleotide array study,. J Cell Physiol 2005;204:522–31.

[48] Huang JX, Kaeslin G, Ranall MV, Blaskovich MA, Becker B, Butler MS, Little MH, Lash LH, Cooper MA. Evaluation of biomarkers for in vitro prediction of drug induced nephrotoxicity: comparison of HK-2, immortalized human proximal tubule epithelial, and primary cultures of human proximal tubular cells. Pharma Res Per 2015;3(3):1–14.

[49] DiMasi JA. The value of improving the productivity of the drug development process faster times and better decisions. Pharmacoeconomics 2002;20:1–10.

[50] Pound P, Ebrahim S, Sandercock P, Bracken MB, Roberts I. Where is the evidence that animal research benefits humans. BMJ 2004;328:7514–7.

[51] Greek R, Pippus A, Hansen LA. The Nuremberg Code subverts human health and safety by requiring animal modeling. BMC Med Ethics 2012;13:16.

[52] Bhattacharya S, Shoda LKM, Zhang Q, Woods CG, Howell BA, Siler SQ, Woodhead JL, Yang Y, McMullen P, Watkins PB, Andersen ME. Modeling drug-and chemical-induced hepatotoxicity with systems biology approaches. Front Physiol 2012;3:462.

[53] Woodhead JL, Howell BA, Yang Y, Harrill AH, Clewell III HJ, Andersen ME, Siler SQ, Watkins PB. An analysis of N-acetylcysteine treatment for acetaminophen overdose using a systems model of drug-induced liver injury. J Pharmacol Exp Ther 2012;342:529–40.

[54] Shoda LKM, Woodhead JL, Siler SQ, Watkins PB, Howell BA. Linking physiology to toxicity using DILIsym(®), a mechanistic mathematical model of drug-induced liver injury. Biopharm Drug Dispos 2014;35:33–49.

[55] Woodhead JL, Brock WJ, Roth SE, Shoaf SE, Brouwer KLR, Church R, Grammatopoulos TN, Stiles L, Siler SQ, Howell BA, Mosedale M, Watkins PB, Shoda LKM. Application of a mechanistic model to evaluate putative mechanisms of tolvaptan drug-induced liver injury and identify patient susceptibility factors. Toxicol Sci 2016:1–14.

Chapter 8

Role of Molecular Chaperone Network in Understanding In Vitro Proteotoxicity

Tulika Srivastava, Sandeep K. Sharma, Smriti Priya
CSIR-Indian Institute of Toxicology Research, Lucknow, India

INTRODUCTION

Proteins are essential parts of organisms like other biological macromolecules such as polysaccharides and nucleic acids and participate in virtually every cellular process. It has been estimated that average-sized bacteria contain about 2 million protein molecules per cell (e.g., *Escherichia coli* and *Staphylococcus aureus*), by contrast, eukaryotic cells such as yeast cells contain about 50 million proteins and human cells on the order of 1–3 billion [1]. Protein metabolism or homeostasis in the cell is maintained by a cluster of processes, which are more than protein quality control; protein homeostasis encompasses RNA metabolism and processing, protein synthesis, folding, translocation, assembly/disassembly and clearance [2]. Protein homeostasis is a process that governs the "life of proteins in cell" [3]. To undertake these roles, most proteins fold into a specific three-dimensional structure, which is largely determined by their amino acids sequences [4]. Thus the cell has to maintain proper protein homeostasis, or proteostasis, and face the intrinsic and environmental stressors.

Interplay between proteostasis network components maintains the long-term health of the cell. Proteostasis network is highly complex and integrated and defects in any one branch trigger the malfunction of entire network and result in numerous metabolic, oncological, cardiovascular, and neurodegenerative diseases [5]. Proteostasis network components also play critical roles in governing the life of the cell and disturbed protein homeostasis leads to abnormal protein activities, misfolding, and accumulation of selected proteins as proteotoxic insoluble inclusions or aggregates. The imbalance in protein homeostasis is caused by age, genetic, epigenetic, physiological, and environmental stressors and leads to proteotoxicity [2]. The first definition of proteotoxicity was given

In Vitro Toxicology. https://doi.org/10.1016/B978-0-12-804667-8.00008-0

TABLE 8.1 Proteotoxicity and Resulted Disorders (Proteopathies)

Protein Aberration	Related Disorder	References
Amyloid beta	Alzheimer's disease (AD), cerebral β-amyloid angiopathy, retinal ganglion cell degeneration	[10,11]
α-Synuclein	Parkinson's disease (PD) and multiple synucleinopathies	[12,13]
Hyperphosphorylated Tau protein	Fronto-temporal dementia with parkinsonism	[14]
Prion proteins	Multiple Prion diseases, Kuru, Creutzfeldt–Jakob, Gerstmann–Sträussler–Scheinker syndrome (GSS) disease, fatal familial insomnia	[15,16]
Islet amyloid polypeptide (and amylin)	Type II diabetes	[17,18]
Transthyretin	Familial amyloidotic neuropathy, senile systemic amyloidosis	[19]
Superoxide dismutase	Amyotrophic lateral sclerosis	[20,21]
Huntingtin with polyglutamine expansion	Huntington's disease (HD)	[22]

by its analogy with the term genotoxicity (damage to DNA by chemical and physical agents), the term proteotoxicity should be useful to describe damage to proteins caused by chemical and physical agents [6]. Stress provides a favorable environment for protein misfolding and a number of neurological and non-neurological disorders are reported because of abnormal protein homeostasis and the accumulation of toxic proteins (Table 8.1) spanning over all tissues of the body significantly highlighting the vast effects of proteotoxicity. The characteristic pattern of misfolded protein accumulation is also used as a hallmark feature of disease pathology. A group of roughly 20 protein deposition diseases, usually referred to as amyloidoses, are characterized by the presence of fibrillar aggregates deposits found as intracellular inclusions or extracellular plaques (amyloid) whose main constituent is a specific peptide or protein [7].

BALANCE BETWEEN PROTEIN FOLDING AND MISFOLDING: MISFOLDED PROTEINS ARE TOXIC SPECIES

A polypeptide encounters number of transformations to fold into native structure and failing to do so results in a misfolded protein structure. A polypeptide has all information to fold into a three-dimensional structure in its amino acid sequence, however, it can fall into one of the many conformational pathways to attain a final native structure with energy minima [4,8]. Environmental stress factors influence the native folding of a polypeptide and cause improper intermolecular interactions, which impair the folding process. The partially folded polypeptide trapped into low-energy nonnative misfolded conformations with exposed hydrophobic residues are prone to sticking with each other [9] and eventually leads to aggregation as observed in amyloid fibrils. The amyloid fibrils are more stable species and even low in energy than the native protein itself [8].

MECHANISMS OF CELLULAR PROTEOTOXICITY

It is now well established that the misfolded forms of proteins are the molecular basis of protein aggregation into amyloid structures [23]. In protein aggregation mechanism, oligomer formation is a key event and is the rate-limiting step responsible for the lag phase in aggregation kinetics [24,25]. Also, oligomers are now established as the pathogenic species associated with the formation of amyloids in diseases [26,27]. Although the exact cause of protein misfolding and initiation of protein aggregation is not known, the "nucleation growth" mechanism is one of the most widely accepted mechanism proposed for the assembly of monomers into oligomers [28,29]. The mechanism explores that monomers convert into a nucleus through a thermodynamically unfavorable process occurring in the lag phase of amyloid aggregation kinetics. The misfolded protein monomers self-assemble into oligomers with different morphologies and structures, which undergoes a structural reorganization into an amyloid-like oligomer and acts as a nucleus. The nucleus rapidly triggers aggregation and leads to the formation of higher order oligomers and fibrils (Fig. 8.1) [30].

The quality control system can adapt to the severity of protein misfolding and damage induced as a stress responses. However, when the generation of misfolded proteins exceeds the refolding or degradation capacity of a cell, protein aggregates accumulate and exhaust the cellular protein quality control system. Various kinds of stress conditions have been identified, which leads to protein misfolding at first place (Fig. 8.1). Most common among all are the

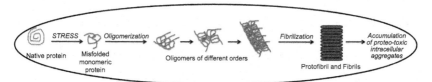

FIGURE 8.1 Protein aggregation pathway. A generalized aggregation pathway adopted by proteins upon exposure to environmental or genetic stress. The native monomeric protein misfolds exposing hydrophobic residues, which have the tendency to stick together and form small oligomers of different morphologies. These oligomers further aggregate to form larger fibrils and aggregates. The toxicity of each species is differently exerted in vitro toward various cellular processes.

mutations that result in the sustained tendency of the affected proteins to misfold and aggregate. The defects in protein biogenesis are translational errors, which lead to incorporation of amino acids and assembly defects of protein complexes resulting into the accumulation of protein species prone to aggregation [31,32]. Environmental stress conditions such as heat and oxidative stress lead to the bulk unfolding of cellular proteins. Apart from these, oxidative stress can cause irreversible protein modifications by reactive oxygen species and carbonyl derivates can be generated by a direct oxidative modification of Pro, Arg, Lys, and Thr residues. These irreversible modifications can then lead to misfolding and eventually aggregation. Protein aggregation in cells is also observed during ageing, although at a slower pace [31]. Misfolded proteins may be deposited as microscopically visible inclusion bodies or plaques within cells and interfere in a wide range of cellular processes to elicit cellular toxicity. Primarily, toxicity arises due to inhibition of synaptic function; loss of synapses leading to disruption of neuronal functions; sequestration of critical cellular chaperones and vital transcription factors by misfolded proteins; interference with numerous signal transduction pathways; alteration of calcium homeostasis; release of free radicals and consequent oxidative damage; dysfunction of the protein degradation pathway through the ubiquitin proteasome system (UPS); and/or induction of cell death proteases leading to programmed cell death [33,34].

CELLULAR DEFENSE AGAINST PROTEOTOXICITY: MOLECULAR CHAPERONES

Cells have a remarkable capacity to buffer changes in the highly dynamic intracellular environment on exposure to the proteotoxic stress. Given the fact that nearly 20%–30% of all proteins in mammalian cells are intrinsically disordered [35] and can engage in promiscuous molecular interactions, cells considerably invest in a number of protein quality control factors to prevent protein misfolding and aggregation. Among different components of protein quality control machinery, molecular chaperones and degradation machinery of proteostasis plays an essential role. Molecular chaperones are prominent modulators of protein homeostasis and prevent protein toxicity at first place. The major functions of molecular chaperones are (1) de novo folding of newly synthesized proteins, (2) preventing aggregation of unfolded proteins, (3) removing misfolded proteins by degradation, and (4) resolubilizing protein aggregates for subsequent refolding or degradation. In unstressed cells, the molecular chaperones play a central role in physiological protein homeostasis. They may regulate structural transitions between "native" and "nonnative" conformations of the proteins [36,37]. In stressed cells, the molecular chaperones become a primary line of cellular defenses against stress-induced protein misfolding and aggregation events [38] that otherwise become increasingly toxic by compromising with the stability of other proteins and the integrity of membranes [39].

The molecular chaperones are classified into five families of highly conserved proteins: the Hsp100s (ClpB), the Hsp90s (HtpG), the Hsp70/Hsp110 (DnaK), Hsp60/CCTs (GroEL), and the α-crystallin-containing domain generally called the "small Hsps" (IbpA/B) (*Escherichia coli* orthologs shown in parentheses). Each family is comprised of multiple members that share sequence homology, have common functional domains, expressed in different subcellular compartments and at different levels in the tissues [40,41]. All molecular chaperones can bind the misfolding intermediates and prevent protein misfolding and aggregation [31,42,43]. All chaperone families have the ability to screen for misfolded proteins with exposed hydrophobic residues to the aqueous phase and are prone to self-association forming stable inactive aggregates [37,44,45]. The complementary mechanisms that optimize the concerted action of each chaperone within the larger network of molecular chaperones and proteases ultimately control the cellular protein homeostasis [46–48]. All major classes of molecular chaperones are ATPases, which need ATP-driven energy to refold the low-energy misfolded or alternatively folded polypeptide substrates bound to them. In addition to this, several reports on ATP-independent action of chaperones such as GroEL in the refolding of misfolded proteins and sHsps (heat shock proteins) in the prevention of aggregation [9,49].

ROLE IN PROTEOTOXICITY: MOLECULAR CHAPERONES ANTAGONIZES THE ACCUMULATION OF PROTEOTOXIC PROTEINS

Chaperones act by multiple means to maintain the solubility of disease-causing toxic proteins. Members of the chaperone family, besides participating in de novo folding of nascent chains, are also involved in the refolding of misfolded protein species and resolubilizing aggregates, which is a prominent activity for the removal of toxic protein aggregates. Chaperone-mediated solubilization can occur in the initial stages of the amyloid assembly pathway by preventing the self-association of misfolded proteins into toxic and oligomeric intermediates. Alternatively, chaperones also act later in the assembly pathway to dismantle amyloid-like aggregates and resolubilize the aggregated proteins. Thus, chaperone intervention can either maintain or generate monomers of the protein, which can be properly refolded or marked for degradation by the ubiquitin-proteasome system or autophagy pathways. This reduces the flux of misfolded or aggregated proteins through different stages of the amyloid assembly pathway and decreases the cellular concentration of proteotoxic assembly intermediates. Different families of chaperones are involved in misfolding repair network at various organelles and sites and target different substrates.

Hsp70 (DnaK)–Hsp40 (DnaJ)–Hsp110 (GrpE) system (also called KJE system in *E. coli*) is involved in reactivation of stress denatured proteins both in vitro and in vivo [50–52]. The Hsp70 is a conserved family of proteins from bacteria to eukaryotes having a critical role in various cellular processes beyond defense against stress damage, such as de novo protein folding, protein translocation, vesicular trafficking, and cellular signaling [52]. Hsp70 collaborates with a J domain cochaperone Hsp40 (DnaJ, CpbA, or DjlA in *E. coli*) that targets the chaperone onto its misfolded protein substrates and together with a nucleotide exchange factor (NEF, GrpE in *E. coli*) controls the coupling between ATP consumption and the unfolding work of the Hsp70 chaperone (Fig. 8.2) [52–54]. In protein aggregation disorders, such as Parkinson's disease (PD), Hsp70 system plays an important and central role in regulating toxic protein aggregation at the first place and maintains the proteome integrity and cell viability [55].

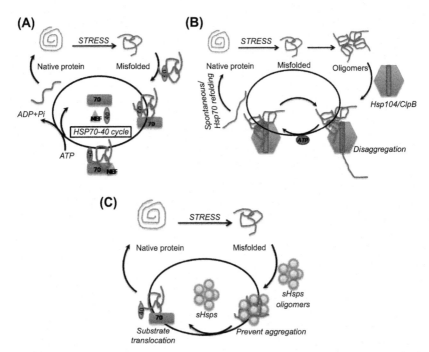

FIGURE 8.2 **Cellular mechanism of major molecular chaperone networks.** (A) Heat shock protein 70 (Hsp70) chaperone system interacts with misfolded monomeric substrates. In the cycle, Hsp40 first interacts with substrate protein and takes to the ATP-bound form of Hsp70, where substrate-binding domain (SBD) interacts with the unfolded polypeptide and upon the hydrolysis of ATP to ADP stimulated by Hsp40, a more stable interaction between the Hsp70 and the substrate is formed. An NEF interacts with the complex exchange ADP for ATP, substrate, and NEF are released from Hsp70 leading to folding into native form. Failed folding leads to the misfolded substrate and enters another round of the Hsp70 reaction cycle. (B) Hsp104 (ClpB) is targeted for small oligomeric misfolded proteins disaggregation and generates the mechanical force to pull apart large aggregates, which could eventually either resolubilize medium size aggregates or misfolded polypeptides. The smaller aggregates are either transported to the DnaK chaperone system directly or enter the cavity of the ClpB hexamer again. (C) Small heat shock protein (sHsp) passively protect cells against potentially deleterious aggregates by remodeling their oligomers. The hetero-oligomeric sSHP:substrate complexes can be recovered by disaggregating Hsp100/Hsp70 chaperone networks.

The ability of Hsp70 and Hsp40 to suppress the assembly of toxic proteins into amyloid-like aggregates has been well documented in numerous in vitro model systems [56–58]. Elevating intracellular pools of Hsp70 as well as Hsp40 reduced toxic protein aggregate formation and suppressed toxicity in cultured cells [57]. Overexpression of the yeast Hsp70, Ssa1, or the Hsp40, Ydj1, inhibits the formation of large, detergent-insoluble aggregates by an expanded-polyQ protein containing the N-terminal sequence of the Huntingtin (Htt) protein, the causative gene for Huntington disease [59]. Ydj1 also reduced the aggregation of the Q/N-rich prion domain of the yeast prion, correlated with its ability to suppress prion toxicity [60]. Elevated expression of yeast Hsp40, Sis1, reduced the aggregation of expanded-polyQ Htt protein and suppressed toxicity in yeast [61]. Hsp70 and Hsp40 have also been shown to suppress the aggregation and toxicity of numerous disease-causing proteins such as mutant superoxide dismutase 1 [62], the causative gene for familial amyotrophic lateral sclerosis, amylin aggregation [63], and α-synuclein [64] responsible for PD. Thus, the ability of Hsp70 and Hsp40 to prevent assembly of amyloid-like aggregates and suppress cytotoxicity extends to numerous disease-causing proteins in a variety of model organisms [54]. It is not only the KJE system that is capable of refolding stress denatured proteins but also other chaperones such as Hsp90 in cooperation with Hsp70 participate in the recovery of cells by promoting either refolding or degradation of misfolded proteins [65,66].

Because protein misfolding leads to aggregation, the evolved cellular mechanism is there to disaggregate and efficiently resolubilize and refold proteins by chaperones in subsequent steps. The phenomenon of disaggregation was first reported in *Saccaromyces cerevisiae* where heat denatured proteins can be efficiently reactivated by the cooperative action of Hsp70–Hsp40 and the oligomeric, ring forming AAA+ ATPase chaperone Hsp104 [67]. Soon after the discovery of Hsp104 disaggregase in yeast, an orthologous protein called ClpB was shown to possess disaggregation activity in *E. coli* and in chloroplasts and mitochondria of higher eukaryotes [51,68]. The mechanism of protein disaggregation by this bichaperone system involves an essential activity of the Hsp70 system during the initial phases of the process. The binding of Hsp70 and J proteins to aggregates restricts the access of proteases to the aggregates and allows the transfer of aggregated polypeptides to the substrate-processing pore of Clpb or Hsp104 (Fig. 8.2). Though the precise mechanism of disaggregation is still elusive, some evidence suggests that Hsp70-Hsp40 remodels protein aggregates and allows the transfer of aggregated polypeptides to the substrate processing pore of Hsp104 [69]. Hsp104, in an ATP dependent process, then exerts a threading or pulling activity to facilitate extraction of the misfolded polypeptides from the aggregates [31]. Once inside the Hsp104 cylinder, the polypeptide is disentangled to a soluble form. Upon exit from the Hsp104, the solubilized protein can re-enter chaperone-mediated refolding cycles. Reports about a disaggregation activity in mammalian cells [70], *Caenorhabditis elegans* [71] and in a cell-free system with Hsp110 (Apg-2) have been identified [72].

Hsp104 possesses a powerful amyloid-remodeling activity and couples ATP hydrolysis to the rapid deconstruction of amyloid fibers composed of toxic proteins such as prion proteins Sup35 and Ure2 [73,74]. Overexpression of Hsp104 is sufficient to eliminate Sup35 prions [75] and Hsp104 can alone dissolute amyloid structure. In the presence of Hsp70 and Hsp40, activity of Hsp104 against various amyloids is improved in vitro and in vivo [76–78]. Hsp104 is also able to resolve preamyloid oligomers of Sup35, which adopt a generic conformation shared by many disease-associated amyloidogenic proteins [73]. Hsp104 or ClpB or its homologs are not found in mammalian systems to disaggregate the proteotoxic aggregates or oligomers. However, recently Hsp110 mammalian disaggregase has been reported to disaggregate the α-synuclein, amyloid β, and prion aggregates formed in PD, Alzheimer's disease (AD), and prion diseases, respectively [79]. Hsp110 is structurally and functionally related to Hsp70 and share a conserved nucleotide-binding domain [80]. The disaggregation activity of Hsp110, which can dis-assemble the amyloids and protein aggregates to nontoxic monomers at physiological rates in an ATP-dependent manner [81,82]. In *C. elegans*, depletion of Hsp110 family proteins by RNA interference (RNAi) [70] or Hsp105 knockout mouse cells shows severe defects in protein aggregation clearance [83]. Remodeling of detergent insoluble toxic protein aggregates [84] is an extraordinary feature of Hsp110–Hsp70–Hsp40 and has vast implications in modulating toxic protein aggregates and maintaining proper cellular homeostasis.

The sHsps have also been reported to participate in disaggregation process by collaborating with the Hsp70–Hsp40–Hsp104 mediated protein disaggregation process and Hsp27-like sHsps can directly interact and bind to the aggregating proteins (Fig. 8.2). This interaction of sHsps with aggregates may shield the nonnative surfaces and hence serves to protect the misfolded proteins from proteases. In an ATP-independent process, an aspect that is remarkably different from other chaperones, sHsps induce changes in the physiochemical properties of misfolded polypeptides in the aggregates. This presumably facilitates the disaggregation machinery to effectively separate and unfold the individual polypeptides in the aggregates [85].

Apart from molecular chaperone network, the degradation machinery is also a part of the protein quality network and comprises of UPS and autophagy components to ensure the timely removal of nonfunctional unrefoldable misfolded proteins. Because both chaperone-mediated refolding and degradation are ATP-dependent processes from a thermodynamic perspective refolding is preferred over degradation. This kind of cellular triage decision whether to refold or to degrade proteins is critically dependent on cochaperones like CHIP (C-terminal Hsp70 Interacting protein) and E3 ubiquitin ligases. Cells may also sequester deleterious protein species to specific cellular sites in order to avoid toxic effects of protein aggregation. Such deposition sites are aggresomes [86] or juxtanuclear quality-control compartment and insoluble protein deposit sites in yeast and higher eukaryotes [87]. This organization would probably allow cells to facilitate the efficient clearance of aggregate in subsequent steps of either disaggregation or degradation.

THE CELLULAR DEGRADATION MECHANISM IN PROTEOTOXICITY

When molecular chaperones fail to mitigate the toxicity of misfolded proteins by refolding them into native forms it instigates the accumulation of misfolded proteins in the cells. To counter these adverse conditions, cells have evolved proteasome and autophagy, the two proteolytic systems with a fundamental role in protein quality control to reduce the damage at the level of proteins. Damaged, misfolded, aggregated, or unnecessary proteins are degraded or removed by the proteasome or through autophagy-lysosome pathway (ALP) to prevent failure of protein quality control and maintenance of cellular homeostasis [88]. Alterations that reduce their functionality favor accumulation of toxic protein aggregates that alter cellular trafficking and trigger cell death as reported in numerous protein conformational disorders and a general biochemical mechanism underlying such disorders.

THE UBIQUITIN PROTEASOME SYSTEM

The UPS is the primary key component of the proteostasis network to terminate damaged proteins selectively in eukaryotic cells. The substrates of UPS are misfolded proteins that are recognized by their exposed hydrophobic residues on the surface and improper disulfide bonds generated because of misfolding [89]. A large number of short-lived proteins that are ineffective in the cytoplasm, nucleus, endoplasmic reticulum (ER), and other cell compartments are also degraded by UPS. This is a complex mechanism where proteins are first targeted by the molecular chaperones, tagged by the ubiquitination machinery, and further recognized, unfolded, and proteolyzed by the proteasome. Importance of UPS in protein aggregation diseases is highlighted by the inhibition of UPS with lactasystine or MG132 like inhibitors, which resulted in aberrant protein aggregates accumulated in the cell and resembling protein aggregation disorders [90].

The structural components of UPS are 26S proteasome composed of the central barrel-shaped catalytic 20S core and the two flanking 19S regulatory complexes. The 20S proteasome can degrade substrates either alone or in association with regulatory particles to form a complex, the 26S proteasome, which specifically recognizes ubiquitin (Ub)-tagged proteins [91,92]. The crystal structure of proteome shows presence of the core 20S particle with a barrel-shaped structure composed of 28 subunits, which are assembled into four seven-member rings stacked back to back as αββα [92a]. The two outer rings constitute of seven α-subunits (α1 to α7), whereas the two inner rings constitute of seven β-subunits (β1–β7) that contain the proteolytic active sites positioned on the interior face of the cylinder, having chymotryptic (β5 subunit), tryptic (β2 subunit), and caspase-like activities (β1 subunit). The 19S regulatory particles on the ends of the 20S proteasome constitute of at least 18 subunits. Its base contains six homologous ATPases subunits (rpt) in a ring and three non-ATPase subunits (rpn) adjoin the outer ring of the 20S particle.

UPS-mediated degradation of substrate protein is comprised of ubiquitination, conjugation, and ligation, deubiquitination, and ultimately proteolysis. This whole process involves ATP-dependent enzymatic action with ubiquitin (Ub) activating enzyme E1, Ub conjugating enzyme E2, and Ub ligating enzymesE3. UPS-mediated degradation involves covalent attachment of multiple Ub molecules to the target protein and subsequent degradation of the Ub tagged protein. Conjugation of Ub to the substrate also occurs via a separate mechanism where Ub is activated in its C-terminal Gly by the Ub-activating enzyme E1. After activation E2 transfers the Ub to E3, which is ligase and bound to substrate protein as well. E3 Ub ligase catalyzes the last step of covalent attachment of Ub to the substrate. Further, a polyubiquitin chain is synthesized by transfer of additional Ub moieties to Lys48 of the previously conjugated Ub molecule [93]. The Ub chain serves as a recognition marker for the protease. Recognition of substrates is done with different E3 ligases either by directly binding to E3 or meditated by the posttranslational modifications like phosphorylation or transfer mediated by molecular chaperones. Among E1 and E2, only single E1 activates Ub required for all modifications and transfer Ub to several species of E2 enzymes, and each E2 acts with either one or several E3s, although only a few E3s have been identified so far.

Rpn10/Mcb1, one of the subunit of the 19S complex, is ubiquitin-binding subunit, which recognizes ubiquitinated proteins and other substrates of the proteasome [94]. The ATPases present on the 19S regulatory particles bind to the poly-ubiquitylated proteins and with the process of ATP hydrolysis it unfolds and translocates the targeted protein into the 20S particle for degradation. As soon as the protein binds to the 19S particle, the polyubiquitin chain is removed and disassembled by deubiquitinating enzymes (DUBs) and the linearized substrate protein is translocated into the catalytic 20S core chamber and cleaved by its proteolytic sites resulting in the formation of small peptides ranging from 3 to 25 residues in length. Peptides that are released by the proteasome are immediately degraded into amino acids by the abundant cytosolic endopeptidases and aminopeptidases [90,95]. In cells, the misfolded proteins bound to molecular chaperones are targeted by specific chaperone class proteins and translocated to UPS machinery (Fig. 8.3). In mammals, The U-box domain proteins, such as Ub fusion degradation protein 2 (UFD2) and CHIP, interact with Hsp70 and guide the translocation of the misfolded cytosolic proteins to

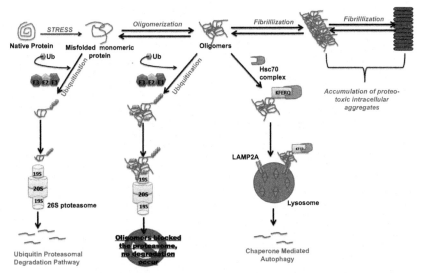

FIGURE 8.3 **Cellular degradation pathways to mitigate in vitro toxicity of proteotoxic misfolded and aggregated proteins.** (A) Ubiquitin proteasomal system is the first system in cell to remove misfolded proteins by tagging them with Ub tags and target for proteasomal degradation. Although the size of the proteasomal cavity cannot process larger accumulated aggregates and inhibited by them. (B) The chaperone mediated autophagy is selective for substrates and degrade only tagged proteotoxic misfolded oligomers, which are targeted by Hsc70 complex with specific peptide tag. The proteotoxic species, which escape from theses cellular defense pathways, are targeted to specific sites for deposition and accumulation, which are probably less harmful for cellular processes. The severity of these toxic accumulations is evident in the form of various diseases progressions.

cellular UPS. The E3 ligase CHIP mediates the transfer of misfolded protein substrate from chaperone to UPS and also serves as a prominent link between two important quality control networks of the cell system. CHIP contains three TPR (tetratricopeptide repeat) domains at its amino terminus, which are used for binding to Hsp70 as well as Hsp90 and a U box domain at its C-terminus, which plays a role in targeting proteins for ubiquitylation and accompanied with proteasome-dependent degradation [96].

AUTOPHAGY

Initial studies projected that the UPS is the main proteolytic system responsible for the degradation of aggregates but more recent data suggest that the lysosome may be another pathway for aggregate removal by ALP and is an important component of protein quality control. The autophagic pathways, i.e., macroautophagy, chaperone-mediated autophagy (CMA), and microautophagy, involve the translocation of aggregates into lysosomes, followed by degradation and the reutilization of the resultant monomer. In many studies, it was found that inhibition of UPS machinery in dopaminergic neurons has been found to induce autophagy via a mechanism requiring p53 [97]. It was shown that due to proteasome inhibition the levels of p53 are increased and multiple pathways have been elucidated through which increase in p53 is proposed to upregulate autophagy. The protective effect of the compensatory upregulation of autophagy in proteasome-inhibited cells has also been suggested to be dependent on histone deacetylase 6 (HDAC6), a microtubule-associated deacetylase that interacts with poly-ubiquitinated proteins, and found to be an essential mechanistic link in between autophagy and UPS [98,99]. However, the role for HDAC6 in this process is not thought to be via signaling to increase autophagic flux, but rather through ensuring efficient delivery of substrates to the autophagic machinery for degradation. Lysosomal proteolysis was initially considered to be a nonselective system, but presently it has been shown that chaperones and other cargo-recognition molecules such as ubiquitins also decide the degradation of specific proteins by the lysosome [100,101]. Therefore, the proteasome and autophagy might be interlinked by using ubiquitin as a common marker for proteolytic degradation (Fig. 8.3). CMA is a highly selective process and does not involve vesicle formation; however, in this process the substrate proteins cross the lysosomal membrane directly to reach the lysosomal lumen. Only cytosolic proteins having a KFERQ-like amino acid sequence (CMA targeting motif) are recognized by the Hsc70 chaperone or a complex of chaperones and cochaperones [102]. As the substrate recognition occurs, they are translocated one by one to the lysosomal membrane, where interaction with the cytosolic tail of the lysosome-associated membrane protein type 2A, Lamp2a, occurs [103]. The defined coordination between cytosolic chaperones and different regulatory factors at the lysosomal membrane allows lysosomes to perform similar selective removal of proteins through CMA as UPS. Coordination between UPS and CMA maintains healthy cellular protein homeostasis.

SPECIFIC ROLE OF DEGRADATION PATHWAYS IN MITIGATING PROTEOTOXICITY

The toxic misfolded proteins are degraded by cellular protein quality control systems, although some of them are prone to aggregate into β-sheet-enriched oligomers and become resistant to all known proteolytic pathways. The accumulation of protease-resistant misfolded and aggregated proteins is a common biochemical mechanism implicit in protein misfolding disorders, including neurodegenerative diseases such as Huntington's disease (HD), AD, PD, prion diseases, and amyotrophic lateral sclerosis (ALS). The pathogenesis of these neurodegenerative diseases is also partly connected with the downregulation of the degradation systems such as UPS. Furthermore research has shown that impairment of proteasomal activity and reduced autophagic potential increase the neurodegenerative phenotype. We discuss here the specific cases of toxic protein misfolding and its impact as dysfunctional proteostasis causing neurodegenerative disease.

TOXICITY OF AMYLOID BETA IN AD

AD is the most common form of progressive dementia, characterized by cognitive impairment, memory loss, and behavioral abnormalities [104]. Amyloid precursor protein (APP) is physiologically a transmembrane protein with unknown functions. Under stress, mutations and due to other unknown mechanisms, APP undergoes two sequential cleavages and produce amyloid-β (Aβ) 1–40 and Aβ 1–42 peptides. Aβ 1–42 is highly aggregation prone peptide and believed to initiate the protein aggregation. In AD Aβ peptides, hyperphosphorylated tau protein are accumulated as misfolded protein or aggregates, which eventually produce amyloid plaques and neurofibrillar tangles, these are hallmark pathogenic feature of the disease. Under ideal conditions in the cell APP and Aβ can be degraded by the UPS at various steps of processing, from the ER lumen to the plasma membrane. The terminally misfolded APP is removed via ER-associated degradation in which substrates are unfolded, ubiquitinated, retrotranslocated across the ER membrane and degraded by the proteasome. Proteasomal degradation can also occur when APP arrives at the Golgi apparatus, where APP is ubiquitinated through a K63 linkage by unknown E3 ligases stimulated by ubiquilin-1, followed by the preservation of APP without proteasomal degradation. Aβ is prone to misfold and targeted by UPS-dependent protein quality control, which involves the E3 ligase CHIP that facilitate the ubiquitination of misfolded proteins for proteasomal degradation. Initially, such aggregates are subjected to degradation, but acute accumulation of these can cause the downregulation of proteolytic pathways triggering a feed-forward loop that in the end destroy essential proteolytic networks. In affected neurons, as compared to APPs, Ub-conjugated Aβ is not properly degraded through the proteasome. Along with APP and Aβ, accumulation of neurofibrillary tangles constituted primarily of phosphorylated tau also degraded through the E3 ligase CHIP, which performs the ubiquitination of tau (primarily in its phosphorylated form) in association with Hsp70 and Hsp90 [105]. Also, the E2 enzyme Ube2w mediates E3-independent ubiquitination of tau [106]. However, ubiquitinated tau is not a good substrate of the proteasome and thus accumulates as detergent-resistant aggregates leading to the formation of neurofibrillar tangles in the affected neurons in AD. In the process of targeting tau to the proteasome, CHIP also appears to be deposited to neurofibrillar tangles with its substrate and other ubiquitinated proteins. Thus, failure of degradation mechanism results in a toxic accumulation of proteotoxic species and their exaggerated effects on cellular functions.

TOXICITY OF α-SYNUCLEIN IN PD

PD is another most common neurodegenerative movement disorder caused by proteotoxicity. It is characterized by decreased motor ability and the loss of dopaminergic neurons in the substantia nigra pars compacta due to accumulation of aggregated proteins mainly α-syn [12]. α-Synuclein, a presynaptic nerve terminal protein with unknown physiological roles becomes prone to aggregation due to environmental factors like exposure to pesticides [107] and mutations (such as A53T, A30P, and E46K) [108,109]. In normal conditions, phosphorylated α-syn at Ser129 can be easily degraded via the proteasomal pathway in an ubiquitin-independent manner in addition to undergoing dephosphorylation in the cell [110]. The filamentous form of α-syn can interact directly with the 20S core of the proteasome and decrease its proteolytic activity. Several regulatory proteins of the UPS were concerned in the yield of soluble α-syn in the cytosol, including Ub ligases CHIP, SIAH (seven in absentia homolog), MDM2 (mouse double minute 2 homolog), and HRD1 [111]. Wild-type α-syn has been found to be ubiquitinated and degraded by the proteasome, using in vitro assays and cultured neuronal cells under proteasomal inhibition [112]. In the pathogenesis of PD, monomeric and nonfibrillar mutant α-syn oligomers are more cytotoxic than fibrillar aggregates [12]. Formation of Lewis bodies composed of α-syn, Ub, the E3 ligase parkin, and other UPS related proteins in PD brains may be a consequence of cytoprotective responses [111,113]. But in the substantia nigra of PD patients, proteasome misregulation has been observed. In addition to Ub ligases, rare mutations in the DUB enzyme UCH-L1 (ubiquitin carboxy-terminal hydrolase L1) have been associated with familial, early cause of PD [114]. During PD pathogenesis, the activity of UCH-L1 is partly contributed as an E3 ligase, whereby it mediates K63-linked ubiquitination in its dimer form, followed by degradation of soluble α-synuclein via proteasome. Still, at present, the nature of the interaction between α-synuclein and the 26S proteasome remains to be traced.

TOXICITY OF HUNTINGTIN PROTEIN IN HD

HD is an autosomal dominant neurodegenerative disorder that affects muscle coordination and leads to progressive cognitive decline and psychosis. This progressive neurodegenerative disease is caused by the aggregation of mutant huntingtin (mHTT) proteins. The wild-type huntingtin protein (HTT) contains a stretch of the glutamine residue, called polyQ tract, which is encoded by a repeat of the codon CAG within exon 1 of the HTT gene. In affected individuals, the CAG repeat expands to >35 in number, giving rise to the elongated polyQ tract of mHTT proteins that are prone to aggregation and are toxic to the neurons [115]. mHTT is a poor substrate for all known proteolytic pathways, including UPS, CMA, and macroautophagy, however, why it is not recognized by any of the proteolytic pathway is unknown. Proteasome activity is found downregulated in brains from HD patients and mice models [116]. mHTT inclusions are constituted of subunits of the UPS, such as Ub and ubiquitinated HTT, as mHTT species can be initially tagged with Ub [111]. The exact mechanisms due to toxicity of inclusion bodies are still unknown but a downregulation of the proteasomal function resulting in proteostasis dysfunction and cellular death has been proposed. It has been suggested that the aggregation of mHTT inclusions is not an outcome of direct proteasomal inhibition but rather result due to the complete inhibition of protein quality control systems in association with the sequestration of molecular chaperones [117].

PRION PROTEINS IN ENCEPHALOPATHIES

Prion diseases, also known as transmissible spongiform encephalopathies, are infectious neurodegenerative disorders in humans and animals. In this disease nervous system including brain are affected resulting in spongiform vacuolation and severe neurodegeneration. In humans, these fatal protein misfolding disorders include kuru, Creutzfeldt–Jakob (CJD) disease, Gerstmann–Sträussler–Scheinker syndrome, fatal familial insomnia, and a new variant, CJD (a human equivalent to bovine spongiform encephalopathy or mad cow disease). The transmissible agent is Scrapie prion protein (PrP^{Sc}), is an abnormally misfolded isoform of the host-encoded cellular prion protein (PrP^C) [118]. PrP^C is a glycosylphosphatidyl inositol-linked glycoprotein and consists of high α-helical content in its structure. A soluble form of misfolded PrP^C is normally degraded by various protein quality control systems, including Ub-dependent ER-associated degradation via proteasome pathway. In mammalian cells, the chaperones GroEL and Hsp104 are found to mediate the conversion of PrP^C into PrP^{Sc} (enriched in β-sheet) in the influence of a small amount of PrP^{Sc} seeds [119]. Despite the implication of chaperones in the number of PrP^C, it appears that PrP^{Sc} cannot be degraded by UPS as it seems to be not a good substrate of the UPS. Moreover, recent studies have shown that binding of PrP^{Sc} to the 20S proteasome without additional processing results in the blocking of substrate entry into the proteolytic chamber, leading to proteasomal failure [120]. PrP^{Sc} may also bind to the external surface of the 20S particle and induce an allosteric stabilization of the closed state of the 20S proteasome [111]. Consistent with these findings, prion diseases are associated with impaired activities of the UPS.

The fine collaboration between protein quality control pathways is important in the maintenance of protein homeostasis, where the inhibition of one system and the activation of the other are essential. Misfolded cellular proteins in the human proteome can be efficiently refolded by molecular chaperones or removed through the cooperative work of the UPS, CMA, and macroautophagy. Complications initiate when aggregation prone disease proteins starts to misfold under stressful environment and become resistant to proteolytic pathways and their β-sheet-enriched folds cannot be unfolded by molecular chaperones to refold again. These pathogenic substrates cannot be properly degraded by the proteasomal pathway and may get stuck within the narrow cylinder of the proteasome, or may not be readily delivered across the lysosomal membrane for CMA. The important questions unresolved in sequestering proteotoxic substrates are the fine cross-talk between different components of PQC (protein quality control) and their regulation and complexity added by specific components of cellular compartments. These connections if resolved could lead to therapeutic wonders in the area of protein aggregation disorders.

CONCLUSION

It is becoming more evident that in vitro protein misfolding and aggregation is not an uncontrolled and catastrophic process. Instead, evolutionarily conserved mechanism direct protein aggregates to specific cellular sites. This specific localization sets the basis for aggregate removal by chaperones and proteolytic machineries. Both aspects will eliminate the damage load of toxic species, thereby reestablishing protein homeostasis. The diversity of the different protein aggregates that have been discussed above (for example, in terms of their structure and cellular localization and related diseases) might have substantial differences in the mechanisms of elimination involved. Understanding the molecular processes regulating this quality control aspect also explains the pathological aspects of various diseases associated with protein aggregation and offer new targets for therapeutic intervention.

Currently, there are no treatments for protein conformational diseases for strengthening cellular proteostasis. Perhaps the most urgent need is for the development of therapeutic approaches and to stop or reverse the progression of proteotoxicity in various disorders. The current use of L-dopamine (L-dopa), a precursor of dopamine, for PD, to improve motor function and patient quality of life has a limited long-term efficacy due to motor complications and drug-induced dyskinesia. Likewise riluzole, the only drug approved for ALS, and the antipsychotics and neuroleptics used in HD only have modest beneficial effects to extend patient survival. There is an urgent need to develop novel therapeutic strategies to treat diseases associated with protein misfolding. The modulation of the proteostasis network through small molecules, which can suppress misfolding and/or aggregation-associated toxicity in various cell or animal model systems of disease, is an interesting target and gained scientific attention in recent years. Compounds in this category are currently in clinical trials for the treatment of some loss- and gain-of-function disorders, respectively. Another approach is represented by small molecules that modulate the activity of specific chaperones such as Hsp90 and Hsp70, which have essential roles in proteostasis. Targeting defense mechanisms is a better approach compared to inhibiting protein aggregation pathway as rest of the species on the pathway may get accumulated with more deleterious effects. Therefore, there is great hope that better understanding of quality control pathways will lead to rational therapeutics to mitigate proteotoxicity and related disorders.

REFERENCES

[1] Milo R. What is the total number of protein molecules per cell volume? A call to rethink some published values. Bioessays 2013;35(12):1050–5.

[2] Balch WE, Morimoto RI, Dillin A, Kelly JW. Adapting proteostasis for disease intervention. Science 2008;319(5865):916–9.

[3] Morimoto RI. Proteotoxic stress and inducible chaperone networks in neurodegenerative disease and aging. Genes Dev 2008;22(11):1427–38.

[4] Anfinsen CB. The formation and stabilization of protein structure. Biochem J 1972;128(4): 737–49.

[5] Dillin A. Andy Dillin: using aging research to probe biology. Interview by Caitlin Sedwick. J Cell Biol 2010;189(4):616–7.

[6] Hightower LE. Heat shock, stress proteins, chaperones, and proteotoxicity. Cell 1991;66(2): 191–7.

[7] Stefani M. Protein misfolding and aggregation: new examples in medicine and biology of the dark side of the protein world. Biochim Biophys Acta 2004;1739(1):5–25.

[8] Wolynes PG. Evolution, energy landscapes and the paradoxes of protein folding. Biochimie 2015;119:218–30.

[9] Priya S, Sharma SK, Goloubinoff P. Molecular chaperones as enzymes that catalytically unfold misfolded polypeptides. FEBS Lett 2013;587(13):1981–7.

[10] Selkoe DJ. Alzheimer's disease: genotypes, phenotypes, and treatments. Science 1997; 275(5300):630–1.

[11] Mattson MP. Pathways towards and away from Alzheimer's disease. Nature 2004;430(7000): 631–9.

[12] Spillantini MG, et al. Alpha-synuclein in Lewy bodies. Nature 1997;388(6645):839–40.

[13] Baba M, et al. Aggregation of alpha-synuclein in Lewy bodies of sporadic Parkinson's disease and dementia with Lewy bodies. Am J Pathol 1998;152(4):879–84.

[14] Hutton M, et al. Association of missense and 5'-splice-site mutations in tau with the inherited dementia FTDP-17. Nature 1998;393(6686):702–5.

[15] Prusiner SB. Prion biology and diseases. Harvey Lect 1991;87:85–114.

[16] Prusiner SB. Prion diseases and the BSE crisis. Science 1997;278(5336):245–51.

[17] Hoppener JW, Ahren B, Lips CJ. Islet amyloid and type 2 diabetes mellitus. N Engl J Med 2000;343(6):411–9.

[18] Hull RL, Westermark GT, Westermark P, Kahn SE. Islet amyloid: a critical entity in the pathogenesis of type 2 diabetes. J Clin Endocrinol Metab 2004;89(8):3629–43.

[19] Sousa MM, Cardoso I, Fernandes R, Guimaraes A, Saraiva MJ. Deposition of transthyretin in early stages of familial amyloidotic polyneuropathy: evidence for toxicity of nonfibrillar aggregates. Am J Pathol 2001;159(6):1993–2000.

[20] Kiernan MC, et al. Amyotrophic lateral sclerosis. Lancet 2011;377(9769):942–55.

[21] Rowland LP, Shneider NA. Amyotrophic lateral sclerosis. N Engl J Med 2001;344(22): 1688–700.

[22] Scherzinger E, et al. Huntingtin-encoded polyglutamine expansions form amyloid-like protein aggregates in vitro and in vivo. Cell 1997;90(3):549–58.

[23] Dobson CM. Protein folding and misfolding. Nature 2003;426(6968):884–90.

[24] Morris AM, Watzky MA, Finke RG. Protein aggregation kinetics, mechanism, and curve-fitting: a review of the literature. Biochim Biophys Acta 2009;1794(3):375–97.

[25] Orte A, et al. Direct characterization of amyloidogenic oligomers by single-molecule fluorescence. Proc Natl Acad Sci USA 2008;105(38):14424–9.

[26] Taylor JP, Hardy J, Fischbeck KH. Toxic proteins in neurodegenerative disease. Science 2002;296(5575):1991–5.

[27] Dobson CM. Protein folding and disease: a view from the first Horizon Symposium. Nat Rev Drug Discov 2003;2(2):154–60.

[28] Buell AK, et al. Solution conditions determine the relative importance of nucleation and growth processes in alpha-synuclein aggregation. Proc Natl Acad Sci USA 2014;111(21):7671–6.

[29] Streets AM, Sourigues Y, Kopito RR, Melki R, Quake SR. Simultaneous measurement of amyloid fibril formation by dynamic light scattering and fluorescence reveals complex aggregation kinetics. PLoS One 2013;8(1):e54541.

[30] Bemporad F, Chiti F. Protein misfolded oligomers: experimental approaches, mechanism of formation, and structure-toxicity relationships. Chem Biol 2012;19(3):315–27.

[31] Tyedmers J, Mogk A, Bukau B. Cellular strategies for controlling protein aggregation. Nat Rev Mol Cell Biol 2010;11(11):777–88.

[32] Valastyan JS, Lindquist S. Mechanisms of protein-folding diseases at a glance. Dis Model Mech 2014;7(1):9–14.

[33] Rao RV, Bredesen DE. Misfolded proteins, endoplasmic reticulum stress and neurodegeneration. Curr Opin Cell Biol 2004;16(6):653–62.

[34] Selkoe DJ. Folding proteins in fatal ways. Nature 2003;426(6968):900–4.

[35] Dunker AK, et al. The unfoldomics decade: an update on intrinsically disordered proteins. BMC Genom 2008;9(Suppl. 2):S1.

[36] Elber R, Karplus M. Multiple conformational states of proteins: a molecular dynamics analysis of myoglobin. Science 1987;235(4786):318–21.

[37] Natalello A, et al. Biophysical characterization of two different stable misfolded monomeric polypeptides that are chaperone-amenable substrates. J Mol Biol 2013;425(7):1158–71.

[38] De Los Rios P, Goloubinoff P. Protein folding: chaperoning protein evolution. Nat Chem Biol 2012;8(3):226–8.

[39] Lashuel HA, Hartley D, Petre BM, Walz T, Lansbury Jr PT. Neurodegenerative disease: amyloid pores from pathogenic mutations. Nature 2002;418(6895):291.

[40] Lindquist S, Craig EA. The heat-shock proteins. Annu Rev Genet 1988;22:631–77.

[41] Gething MJ, Sambrook J. Protein folding in the cell. Nature 1992;355(6355):33–45.

[42] Bukau B, Weissman J, Horwich A. Molecular chaperones and protein quality control. Cell 2006;125(3):443–51.

[43] Hartl FU, Bracher A, Hayer-Hartl M. Molecular chaperones in protein folding and proteostasis. Nature 2011;475(7356):324–32.

[44] Hartl FU, Hayer-Hartl M. Molecular chaperones in the cytosol: from nascent chain to folded protein. Science 2002;295(5561):1852–8.

[45] Hinault MP, Ben-Zvi A, Goloubinoff P. Chaperones and proteases: cellular fold-controlling factors of proteins in neurodegenerative diseases and aging. J Mol Neurosci 2006;30(3):249–65.

[46] Finka A, Mattoo RU, Goloubinoff P. Meta-analysis of heat- and chemically upregulated chaperone genes in plant and human cells. Cell Stress Chaperones 2011;16(1):15–31.

[47] Ranson NA, et al. Allosteric signaling of ATP hydrolysis in GroEL-GroES complexes. Nat Struct Mol Biol 2006;13(2):147–52.

[48] Voisine C, Pedersen JS, Morimoto RI. Chaperone networks: tipping the balance in protein folding diseases. Neurobiol Dis 2010;40(1):12–20.

[49] Priya S, et al. GroEL and CCT are catalytic unfoldases mediating out-of-cage polypeptide refolding without ATP. Proc Natl Acad Sci USA 2013;110(18):7199–204.

[50] Ben-Zvi A, De Los Rios P, Dietler G, Goloubinoff P. Active solubilization and refolding of stable protein aggregates by cooperative unfolding action of individual hsp70 chaperones. J Biol Chem 2004;279(36):37298–303.

[51] Goloubinoff P, Mogk A, Zvi AP, Tomoyasu T, Bukau B. Sequential mechanism of solubilization and refolding of stable protein aggregates by a bichaperone network. Proc Natl Acad Sci USA 1999;96(24):13732–7.

[52] Sharma SK, De los Rios P, Christen P, Lustig A, Goloubinoff P. The kinetic parameters and energy cost of the Hsp70 chaperone as a polypeptide unfoldase. Nat Chem Biol 2010;6(12):914–20.

[53] Sharma SK, Christen P, Goloubinoff P. Disaggregating chaperones: an unfolding story. Curr Protein Pept Sci 2009;10(5):432–46.

[54] Kampinga HH, Craig EA. The HSP70 chaperone machinery: J proteins as drivers of functional specificity. Nat Rev Mol Cell Biol 2010;11(8):579–92.

[55] Labbadia J, et al. Suppression of protein aggregation by chaperone modification of high molecular weight complexes. Brain 2012;135(Pt 4):1180–96.

[56] Ross CA, Poirier MA. Protein aggregation and neurodegenerative disease. Nat Med 2004; (10 Suppl.):S10–7.

[57] Kalia SK, Kalia LV, McLean PJ. Molecular chaperones as rational drug targets for Parkinson's disease therapeutics. CNS Neurol Disord Drug Targets 2010;9(6):741–53.

[58] Hinault MP, et al. Stable alpha-synuclein oligomers strongly inhibit chaperone activity of the Hsp70 system by weak interactions with J-domain co-chaperones. J Biol Chem 2010;285(49):38173–82.

[59] Muchowski PJ, et al. Hsp70 and hsp40 chaperones can inhibit self-assembly of polyglutamine proteins into amyloid-like fibrils. Proc Natl Acad Sci USA 2000;97(14):7841–6.

[60] Summers DW, Douglas PM, Ren HY, Cyr DM. The type I Hsp40 Ydj1 utilizes a farnesyl moiety and zinc finger-like region to suppress prion toxicity. J Biol Chem 2009;284(6): 3628–39.

[61] Gokhale KC, Newnam GP, Sherman MY, Chernoff YO. Modulation of prion-dependent polyglutamine aggregation and toxicity by chaperone proteins in the yeast model. J Biol Chem 2005;280(24):22809–18.

[62] Takeuchi H, et al. Hsp70 and Hsp40 improve neurite outgrowth and suppress intracytoplasmic aggregate formation in cultured neuronal cells expressing mutant SOD1. Brain Res 2002;949(1–2):11–22.

[63] Chien V, et al. The chaperone proteins HSP70, HSP40/DnaJ and GRP78/BiP suppress misfolding and formation of beta-sheet-containing aggregates by human amylin: a potential role for defective chaperone biology in type 2 diabetes. Biochem J 2010;432(1):113–21.

[64] Auluck PK, Chan HY, Trojanowski JQ, Lee VM, Bonini NM. Chaperone suppression of alpha-synuclein toxicity in a Drosophila model for Parkinson's disease. Science 2002;295(5556):865–8.

[65] Pratt WB, Toft DO. Regulation of signaling protein function and trafficking by the hsp90/hsp70-based chaperone machinery. Exp Biol Med (Maywood) 2003;228(2):111–33.

[66] Wegele H, Muller L, Buchner J. Hsp70 and Hsp90–a relay team for protein folding. Rev Physiol Biochem Pharmacol 2004;151:1–44.

[67] Glover JR, Lindquist S. Hsp104, Hsp70, and Hsp40: a novel chaperone system that rescues previously aggregated proteins. Cell 1998;94(1):73–82.

[68] Mogk A, Bukau B, Lutz R, Schumann W. Construction and analysis of hybrid *Escherichia coli-Bacillus subtilis dnaK* genes. J Bacteriol 1999;181(6):1971–4.

[69] Zietkiewicz S, Lewandowska A, Stocki P, Liberek K. Hsp70 chaperone machine remodels protein aggregates at the initial step of Hsp70-Hsp100-dependent disaggregation. J Biol Chem 2006;281(11):7022–9.

[70] Rampelt H, et al. Metazoan Hsp70 machines use Hsp110 to power protein disaggregation. EMBO J 2012;31(21):4221–35.

[71] Bieschke J, Cohen E, Murray A, Dillin A, Kelly JW. A kinetic assessment of the *C. elegans* amyloid disaggregation activity enables uncoupling of disassembly and proteolysis. Protein Sci 2009;18(11):2231–41.

[72] Shorter J. The mammalian disaggregase machinery: Hsp110 synergizes with Hsp70 and Hsp40 to catalyze protein disaggregation and reactivation in a cell-free system. PLoS One 2011;6(10):e26319.

[73] Shorter J, Lindquist S. Destruction or potentiation of different prions catalyzed by similar Hsp104 remodeling activities. Mol Cell 2006;23(3):425–38.

[74] Savistchenko J, Krzewska J, Fay N, Melki R. Molecular chaperones and the assembly of the prion Ure2p in vitro. J Biol Chem 2008;283(23):15732–9.

[75] Chernoff YO, Lindquist SL, Ono B, Inge-Vechtomov SG, Liebman SW. Role of the chaperone protein Hsp104 in propagation of the yeast prion-like factor [psi+]. Science 1995;268(5212):880–4.

[76] Tipton KA, Verges KJ, Weissman JS. In vivo monitoring of the prion replication cycle reveals a critical role for Sis1 in delivering substrates to Hsp104. Mol Cell 2008;32(4):584–91.

[77] Shorter J. Hsp104: a weapon to combat diverse neurodegenerative disorders. Neurosignals 2008;16(1):63–74.

[78] Lo Bianco C, et al. Hsp104 antagonizes alpha-synuclein aggregation and reduces dopaminergic degeneration in a rat model of Parkinson disease. J Clin Invest 2008;118(9):3087–97.

[79] Sadlish H, et al. Hsp110 chaperones regulate prion formation and propagation in *S. cerevisiae* by two discrete activities. PLoS One 2008;3(3):e1763.

[80] Polier S, Dragovic Z, Hartl FU, Bracher A. Structural basis for the cooperation of Hsp70 and Hsp110 chaperones in protein folding. Cell 2008;133(6):1068–79.

[81] Nillegoda NB, Bukau B. Metazoan Hsp70-based protein disaggregases: emergence and mechanisms. Front Mol Biosci 2015;2:57.

[82] Mattoo RU, Sharma SK, Priya S, Finka A, Goloubinoff P. Hsp110 is a bona fide chaperone using ATP to unfold stable misfolded polypeptides and reciprocally collaborate with Hsp70 to solubilize protein aggregates. J Biol Chem 2013;288(29):21399–411.

[83] Yamagishi N, Goto K, Nakagawa S, Saito Y, Hatayama T. Hsp105 reduces the protein aggregation and cytotoxicity by expanded-polyglutamine proteins through the induction of Hsp70. Exp Cell Res 2010;316(15):2424–33.

[84] Gao X, et al. Human Hsp70 disaggregase reverses Parkinson's-linked alpha-synuclein amyloid fibrils. Mol Cell 2015;59(5):781–93.

[85] Mogk A, et al. Refolding of substrates bound to small Hsps relies on a disaggregation reaction mediated most efficiently by ClpB/DnaK. J Biol Chem 2003;278(33):31033–42.

[86] Kopito RR. Aggresomes, inclusion bodies and protein aggregation. Trends Cell Biol 2000;10(12):524–30.

[87] Spokoini R, et al. Confinement to organelle-associated inclusion structures mediates asymmetric inheritance of aggregated protein in budding yeast. Cell Rep 2012;2(4):738–47.

[88] Takalo M, Salminen A, Soininen H, Hiltunen M, Haapasalo A. Protein aggregation and degradation mechanisms in neurodegenerative diseases. Am J Neurodegener Dis 2013;2(1): 1–14.

[89] Ravid T, Hochstrasser M. Diversity of degradation signals in the ubiquitin-proteasome system. Nat Rev 2008;9(9):679–90.

[90] Lecker SH, Goldberg AL, Mitch WE. Protein degradation by the ubiquitin-proteasome pathway in normal and disease states. J Am Soc Nephrol 2006;17(7):1807–19.

[91] Ciechanover A. The ubiquitin-proteasome pathway: on protein death and cell life. EMBO J 1998;17(24):7151–60.

[92] Jung T, Catalgol B, Grune T. The proteasomal system. Mol Asp Med 2009;30(4):191–296.

[92a] Groll M, Ditzel L, Löwe J, Stock D, Bochtler M, Bartunik HD, Huber R. Structure of 20S proteasome from yeast at 2.4 A resolution. Nature 1997;386(6624):463–71.

[93] Peng J, et al. A proteomics approach to understanding protein ubiquitination. Nat Biotechnol 2003;21(8):921–6.

[94] van Nocker S, et al. The multiubiquitin-chain-binding protein Mcb1 is a component of the 26S proteasome in *Saccharomyces cerevisiae* and plays a nonessential, substrate-specific role in protein turnover. Mol Cell Biol 1996;16(11):6020–8.

[95] Saric T, Graef CI, Goldberg AL. Pathway for degradation of peptides generated by proteasomes: a key role for thimet oligopeptidase and other metallopeptidases. J Biol Chem 2004;279(45):46723–32.

[96] Amm I, Sommer T, Wolf DH. Protein quality control and elimination of protein waste: the role of the ubiquitin-proteasome system. Biochim Biophys Acta 2014;1843(1):182–96.

[97] Du Y, et al. An insight into the mechanistic role of p53-mediated autophagy induction in response to proteasomal inhibition-induced neurotoxicity. Autophagy 2009;5(5):663–75.

[98] Pandey UB, et al. HDAC6 rescues neurodegeneration and provides an essential link between autophagy and the UPS. Nature 2007;447(7146):859–63.

[99] Iwata A, Riley BE, Johnston JA, Kopito RR. HDAC6 and microtubules are required for autophagic degradation of aggregated huntingtin. J Biol Chem 2005;280(48):40282–92.

[100] Nixon RA. The role of autophagy in neurodegenerative disease. Nat Med 2013;19(8):983–97.

[101] Wong E, Cuervo AM. Integration of clearance mechanisms: the proteasome and autophagy. Cold Spring Harb Perspect Biol 2010;2(12):a006734.

[102] Arias E, Cuervo AM. Chaperone-mediated autophagy in protein quality control. Curr Opin Cell Biol 2011;23(2):184–9.

[103] Cuervo AM, Dice JF. A receptor for the selective uptake and degradation of proteins by lysosomes. Science 1996;273(5274):501–3.

[104] Selkoe DJ. Alzheimer's disease. Cold Spring Harb Perspect Biol 2011;3(7).

[105] Petrucelli L, et al. CHIP and Hsp70 regulate tau ubiquitination, degradation and aggregation. Hum Mol Genet 2004;13(7):703–14.

[106] Scaglione KM, et al. The ubiquitin-conjugating enzyme (E2) Ube2w ubiquitinates the N terminus of substrates. J Biol Chem 2013;288(26):18784–8.

[107] Betarbet R, et al. Chronic systemic pesticide exposure reproduces features of Parkinson's disease. Nat Neurosci 2000;3(12):1301–6.

[108] Polymeropoulos MH, et al. Mutation in the alpha-synuclein gene identified in families with Parkinson's disease. Science 1997;276(5321):2045–7.

[109] Kruger R, et al. Ala30Pro mutation in the gene encoding alpha-synuclein in Parkinson's disease. Nat Genet 1998;18(2):106–8.

[110] Machiya Y, et al. Phosphorylated alpha-synuclein at Ser-129 is targeted to the proteasome pathway in a ubiquitin-independent manner. J Biol Chem 2010;285(52):40732–44.

[111] Ciechanover A, Kwon YT. Degradation of misfolded proteins in neurodegenerative diseases: therapeutic targets and strategies. Exp Mol Med 2015;47:e147.

[112] Bennett MC, et al. Degradation of alpha-synuclein by proteasome. J Biol Chem 1999;274(48):33855–8.

[113] Schlossmacher MG, et al. Parkin localizes to the Lewy bodies of Parkinson disease and dementia with Lewy bodies. Am J Pathol 2002;160(5):1655–67.

[114] Das C, et al. Structural basis for conformational plasticity of the Parkinson's disease-associated ubiquitin hydrolase UCH-L1. Proc Natl Acad Sci USA 2006;103(12):4675–80.

[115] Williams AJ, Paulson HL. Polyglutamine neurodegeneration: protein misfolding revisited. Trends Neurosci 2008;31(10):521–8.

[116] Waelter S, et al. Accumulation of mutant huntingtin fragments in aggresome-like inclusion bodies as a result of insufficient protein degradation. Mol Biol Cell 2001;12(5):1393–407.

[117] Hipp MS, et al. Indirect inhibition of 26S proteasome activity in a cellular model of Huntington's disease. J Cell Biol 2012;196(5):573–87.

[118] Prusiner SB. Novel proteinaceous infectious particles cause scrapie. Science 1982;216(4542): 136–44.

[119] DebBurman SK, Raymond GJ, Caughey B, Lindquist S. Chaperone-supervised conversion of prion protein to its protease-resistant form. Proc Natl Acad Sci USA 1997;94(25):13938–43.

[120] Deriziotis P, et al. Misfolded PrP impairs the UPS by interaction with the 20S proteasome and inhibition of substrate entry. EMBO J 2011;30(15):3065–77.

Chapter 9

Scientific and Regulatory Considerations in the Development of in Vitro Techniques for Toxicology

Mukul R. Jain, Debdutta Bandyopadhyay, Rajesh Sundar
Zydus Research Centre, Ahmedabad, India

INTRODUCTION

Safety and toxicity are major concerns for any type of product intended for human consumption. In any country, regulatory authorities assume utmost care about the safety of consumers. Thus, toxicity studies play a cardinal role for products such as medicines, agrochemicals, pesticides, and cosmetics. The most reliable toxicity studies are conducted using animal models. However, with rising concerns about animal rights, alternative methods that do not involve animals are being exploited, and in vitro toxicity studies are increasingly important. Importantly, in vitro studies have their own challenges, benefits, limitations, and regulatory dimensions.

WHY IN VITRO TOXICITY?

In 1959, William Russell and Rex Burch suggested that proper experimental design should consider "refining" methods to lessen pain or distress, "reduce" the number of animals necessary without compromising the amount and quality of data, or "replace" the animal model altogether with other alternatives[1]. These three "Rs" lay the foundation for progress related to in vitro toxicity studies. In addition to the three "Rs", other reasons for adopting in vitro assays are described below:

1. **Delineating the mechanism of toxicity**: Although animal studies reveal the outcome of a toxic response, understanding the underlying mechanism is frequently very important. Achieving this objective requires carefully designed in vitro experiments. Knowledge of the mechanism of a given toxicity, in addition to helping chemists modulate the structure, can help answer specific questions that may arise from in vivo toxicity studies.

In Vitro Toxicology. http://dx.doi.org/10.1016/B978-0-12-804667-8.00009-2
165

2. **Cost effectiveness**: In most cases, in vitro toxicity studies are cheaper than whole animal experiments, particularly for screening large numbers of samples.
3. **Time of experiments**: In vitro toxicity studies can usually be completed within a shorter time period than animal studies. However, the exact duration of drug treatment will depend on the nature of the drug, mechanism of drug action, and the period for which the drug is supposed to be taken.

IN VITRO ASSAYS FOR SAFETY AND TOXICITY ASSESSMENT

Knowledge about the mechanism of action and the intended target of a drug is a prerequisite for the successful design of toxicity assays. The endpoint of a given assay can be categorized as pharmacodynamic or toxic, depending on its purpose. For example, a cytotoxicity assay can serve as a toxicity endpoint for antidiabetic drugs but as a pharmacodynamic measure for cancer targets. Similarly, the hepatotoxic property of a drug will be regarded as toxicity if the drug is targeted for breast cancer and pharmacodynamic measure for liver cancer.

Different types of in vitro assay formats have been used for safety, toxicity, and mechanistic studies of chemicals.

Cytotoxicity Assessment

Cytotoxicity is defined as the potential of a compound to cause cell death. The two modes of cell deaths are necrosis and apoptosis. In most cases, cellular necrosis represents in vitro cytotoxicity. However, in certain cases, such as cancer-related cytotoxicity, differentiating between apoptosis and necrosis is extremely valuable. Moreover, determination of the dose-dependence and time kinetics of cell death also sheds light on predicting the actual toxicity. Thus, when developing an assay, special consideration is needed regarding the origin of the cell type, the concentration of cells, the concentration of drug, and the time of treatment.

In addition to the chemical itself, the container or device with which the drug is administered can cause cytotoxicity. Thus cytotoxicity testing is a key element of international standards. The international standards compiled as ISO 10993 [2] and the United States Food and Drug Administration (US FDA) blue book memorandum based on 10993-1 [3] address the critical issue of ensuring device biocompatibility by identifying several types of tests for use in selecting device materials. In vitro studies required for all types of devices, especially cellular toxicity testing, are covered in ISO 10993-5: "Tests for Cytotoxicity—In vitro Methods" [4]. This standard presents a number of test methods designed to evaluate the acute adverse biological effects of extractables from medical device materials.

Use of Cytotoxicity as a Predictor for Acute System Toxicity

Based on a large number of studies on a wide range of chemical entities, a good correlation has been established between cytotoxicity results and actual toxicity observed in intact systems [5]. However, this success greatly depends on the choice of cell line. For example, normal human hepatocytes could have been an excellent choice for testing liver toxicity. However, limited life span, phenotypic instability, restricted accessibility, and high batch-to-batch functional variability complicate the use of these normal cells. By contrast, immortalized hepatoma cells offer unlimited lifespan and a stable phenotype along with easy access. Unfortunately, currently available human hepatoma cell lines are not a good alternative to cultured hepatocytes as they show very limited expression of most drug-metabolizing enzymes. In recent years, recombinant models heterologously expressing cytochrome P450 enzymes in hepatoma cells have also been generated and are widely used in drug metabolism and toxicity evaluations [6].

Considering the pros and cons of both cell types, the use of immortal cell lines representing various organs has become a popular approach in cytotoxicity determination.

Endpoints of Cytotoxicity Studies

The choice of endpoint mainly depends on the specific purpose of the study. These endpoints may include changes in morphology, altered mitochondrial function, change in membrane permeability, and alteration of the cell cycle. Furthermore, for certain cases such as cancer drugs, discriminating apoptosis from necrosis is important. Changes in morphology can be monitored under a microscope, but the most common method for monitoring cell death is based on altered mitochondrial function. Live mitochondria express mitochondrial reductase, which reduces tetrazolium salts to formazan crystals. These crystals are solubilized in DMSO, and a purple color is developed. [3H]-Thymidine incorporation is one of the oldest methods for monitoring cell-cycle alteration and was recently modified by bromo-deoxy-uridine (BrdU) incorporation and detection of the incorporated BrdU by enzyme-linked immunosorbent assay (ELISA). To differentiate apoptosis from necrosis, detection of a DNA ladder and staining by Annexin V are equally popular and effective. Caspase and TUNEL (terminal deoxynucleotidyl transferase dUTP nick end labeling) assays are also attractive methods for the quantitation of apoptosis. Moreover, appropriate caspase selection also provides information about the stage of apoptosis, whereas the TUNEL assay detects only late-stage apoptosis. Recovery from toxicity upon withdrawal of the drug is achieved simply by washing off the drug, followed by incubation in drug-free medium and repeating the same endpoint used to detect toxicity at different time points postwithdrawal.

Applications of Various Conventional and Evolving Techniques for Evaluating Cellular Response Leading to Toxic Insult

Although the techniques mentioned above indicate the endpoint of toxicity, they cannot delineate the mechanism of toxicity. Thus to determine if any cytotoxic pathways are activated at a lower concentration than that required for cell killing, studies of events upstream of the toxicity are a prerequisite. By comparing these values with the circulating concentration of a drug, valuable information about the toxic outcomes of the drug in question can be obtained. Because these pathways originate at the level of genes followed by their transcripts and finally the proteins they encode, detailed genomic, transcriptomic and proteomic studies are becoming extremely popular.

Microarray

Because a toxic response may disturb the expression of a panel of genes, the first and foremost step is to identify these genes. For that purpose, microarray is the method of choice. In microarray, messenger RNA (mRNA) from cells of the treated and untreated sets is converted to their corresponding complimentary DNAs (cDNAs), labeled and subjected to hybridization with cDNAs or polymerase chain reaction (PCR)-amplified fragments of genes immobilized on a solid support [7]. The probes from treated or untreated samples are labeled with green and red fluorescent tags. The differential hybridization of red versus green tagged probes indicates the altered expression. The main challenge is to (1) choose the right type of cell line (e.g., hepatocytes for liver toxicity and cardiomyocytes for heart toxicity), (2) choose the right panel (based on an educated guess), and (3) obtain accurate data. For accuracy, cross-labeling (i.e., switching the label and repeating the experiment) provides reliable results. A successful microarray should identify a set of genes that are linked in one or more pathways. The next step is to study these genes individually.

RT-PCR

Once a small number of genes are identified by microarray, the genes must be verified at the individual gene level. The mRNA(s) of interest are converted to cDNA by reverse transcriptase (RT) and amplified by PCR using specific primer pairs. The amplified fragments are finally sequenced to confirm the identity of the gene. With the development of next-generation sequencing, sequencing has become faster, easier, and very robust.

Proteomics Studies

The last level of alteration leading to cytotoxicity is alterations of the levels of proteins. The levels of proteins are modulated at two levels: transcription and turnover. Thus a parallel connection between the results obtained from RT-PCR and proteomics studies may not necessarily be established in each case. Proteomics studies should not be viewed as downstream of transcriptomics but rather as a parallel approach. The basic components of proteomic studies are (1) separation of proteins by isoelectric focusing in the first dimension; (2) sodium dodecyl sulfate-polyacrylamide gel electrophoresis (SDS-PAGE) in the second dimension; (3) isolation of the proteins with altered expression in treated versus untreated samples; and finally (4) characterization of these proteins by matrix-associated laser-desorption/ionization time of flight.

Benefits of Cytotoxicity Testing

Cytotoxicity testing is a rapid, standardized, sensitive, and inexpensive means to determine whether a material contains significant quantities of biologically harmful components. The high sensitivity of the tests is due to the isolation of the test cells in cultures and the absence of the protective mechanisms that assist cells within the body.

Assessment of Carcinogenicity

This type of toxicity can be caused by direct insult to the genome via mutation or deletion or indirectly by favoring the selection of cells that are destined to be malignant due to spontaneous errors or induced damage.

Genotoxicity Assessment

The major objectives of genotoxicity tests are to identify somatic mutations and germ cell mutations. On the basis of international guidelines, the following tests are performed:

Tests for Gene Mutation in Bacteria

The most widely used short-term test is the Ames test. *Salmonella typhimurium* strains that are deficient in DNA repair and unable to synthesize histidine are used. In the presence of a mutagenic chemical, the defective histidine gene can be mutated back to a functional state (back mutation), resulting in restoration of bacterial growth in medium lacking histidine. The mutant colonies that can make histidine are referred to as "revertants". The Ames test in basic form can detect direct but not indirect genotoxic carcinogens. With the inclusion of a metabolic source, specifically the $9000 \times g$ supernatant (S9) of rat liver homogenate to promote metabolic conversion of the chemical, the Ames test can also detect indirect-acting genotoxic carcinogens. Genetically unique strains of *S. typhimurium* have been developed for determining specific mutational targets. Strains TA100 and TA1535 are able to detect point mutations, whereas strains TA98, TA1537, and TA1538 are able to detect frameshift mutations.

Tests for Gene Mutation in Mammalian Cells

The mouse lymphoma assay is a mutagenicity assay used to determine whether a chemical is capable of inducing mutation in eukaryotic cells. Typically, mouse lymphoma L5178Y cells are used, and the ability of the cell cultures to acquire resistance to trifluorothymidine (the result of forward mutation at the thymidine kinase [TK] locus) is quantified. Another mammalian cell mutation assay, the Chinese hamster ovary test, is also commonly used to assess the potential mutagenicity of chemicals. This assay uses the hypoxanthine–guanine phosphoribosyltransferase (HGPRT) gene as the endpoint. The number of mutant colonies is a measure of the ability of the test chemical to induce a genetic change at the TK or HGPRT locus in these transformed cells.

Tests for Chromosomal Aberration in Mammalian Cells

Organisation for Economic Co-operation and Development (OECD), in their recent updates, has prescribed clear guidelines about tests for chromosomal aberration in mammalian cells [8–13]. They also have provided useful guidelines for cell-based micronucleus tests [14,15] (OECD Guidelines nos. 474 and 487). The most common methods include fluorescent in situ hybridization and the detection of gene mutations at the thymidylate kinase and HPRT (hypoxanthine phosphoribosyltransferase) loci.

Nongenotoxic Carcinogenicity Endpoints

Several chemicals induce hyperplastic responses without causing genomic changes. These agents cause cancers due to a wide variety of mechanisms and can be classified accordingly as follows:

1. Receptor-mediated endocrine modifiers: progestins, medroxyprogesterone acetate (activate progesterone receptor), and 1,4 di-chlorobenzene (activates aryl hydrocarbon receptor)
2. Nonreceptor-mediated endocrine modifiers: 6-propyl-2-thiouracil
3. Persistent cytotoxicity: alcohols [16] and chloroform [17] induce continuous cytotoxicity, thus forcing the organism to regenerate cells
4. Inducers of chronic inflammatory responses: captan [18] and asbestos [19] cause chronic inflammation in the small intestine and lung, respectively
5. Agents causing DNA methylation: methylation of cytosine in the CpG islands of promoters regulates transcription. A well-documented example of this class is bisphenol A, which causes breast and prostate cancer by DNA demethylation upon exposure in utero [20]. By contrast, gynestin increases DNA methylation [21]

Because the agents that modulate nongenotoxic endpoints are capable of exerting their toxic effects long in the future, even to the next generation, considerable attention should be given to these types of changes. Unfortunately, no single method can detect all of these changes, and a holistic approach may help in revealing hidden dangers in apparently safe compounds. Thus, even if a compound is apparently free of genotoxic potential, the possibility of acting as a nongenotoxic agent cannot be eliminated. In view of these considerations, OECD guidelines recommend performing cell transformation assays with high sensitivity for detecting both genotoxic and nongenotoxic carcinogens [22].

Transformation Assays

A variety of in vitro test systems have been developed to assess the carcinogenic potential of chemicals. The C3H/10T½ cell line has been widely used for the transformation assay. These cells are approximately tetraploid, but the chromosome number in the cells varies widely. Consequently, these cells are chromosomally abnormal and have already passed through some of the stages that might be involved in the production of a cancerous cell. Upon plating these cells, they will stop growing when their density is sufficiently high (contact growth inhibition). However, contact inhibition can fail, resulting in cell piling to form a transformed colony. Therefore, following exposure to xenobiotics, this assay assesses carcinogenic potential based on the percentage of colonies that are transformed. Transformation assays using BALB/3T3 and Syrian hamster embryo cells are available for the assessment of the carcinogenic potential of chemicals. The BALB/3T3 transformation assay, however, has not gained full acceptance as a carcinogenic screening assay mainly due to issues regarding its relatively low sensitivity, low reproducibility, and relatively long test period.

Reproductive and Developmental Toxicity Assessment

More than 30 different tests have been proposed as alternatives for in vivo animal models for addressing reproductive and developmental toxicity. These include the use of nonvertebrate species, lower vertebrate embryos, mammalian embryos, micromass culture from mammalian embryos, embryonic stem (ES) cells, and other mammalian cells. Unfortunately, acceptance of these systems to assess risk hazard for humans remains questionable. Some of the assays explored are described below:

1. Cell lines and ES cells: Human embryonic palate mesenchymal cells [23], mouse ovarian tumor cells [24], and neuroblastoma cells [25] have been used to identify developmental toxicity. Unfortunately, results from trials showed high false positives [26]. Tests were also conducted on these cells to assess the potency of a compound to inhibit differentiation and if these cells exhibit different sensitivities than their adult counterparts. Interestingly, the results of these tests were comparable to those of embryotoxicity studies in whole rat embryo culture [27]. However, experimentation with ES cells is somewhat controversial, and the recent progress in converting adult cells to stem cells, i.e., induced pluripotent stem cells [28,29], may circumvent these hurdles.
2. Micromass culture: When cells from chick embryo neural retina [30] and undifferentiated mesenchyme from early embryo limbs [31] are cultured at high density in a small volume, they tend to form foci and exhibit the qualities of undifferentiated cells. The basic principle of using these cells is that any compound with developmental toxicity will prevent the formation of foci. However, because animal body parts are used in these assays, the in vitro nature of these studies is not debatable.
3. In addition to these two systems, embryos from lower species have also been used for developmental toxicity studies [32].

A validation study of three embryotoxicity assays, the rat embryo limb bud micromass assay, the mouse ES cell test (mEST), and the rat whole embryo culture (WEC) test was performed using chemicals that are most predictive of in vivo results [33]. For WEC, rodent or rabbit embryos obtained at an early stage of organogenesis with intact yolk sacs are cultured under conditions mimicking those in utero for 48h and thereafter scored for a number of growth and developmental parameters, enabling the assignment of a developmental score. The mEST uses established stem cell lines and examines the effects of compounds on the viability and differentiation of the cells. In this assay, the cells aggregate and form "embryoid bodies" when grown in hanging-drop cultures. Currently, the rodent WEC test and mEST are receiving greater attention for use as teratogen screens.

Phototoxicity Assessment

Phototoxicity is defined as a toxic response to the application of a substance to the body that is either elicited or increased after subsequent exposure to light or that is induced by skin irradiation after systemic administration of a substance [34]. The most common test used for this purpose is the 3T3 neutral red uptake (NRU) assay. The test is based on a comparison of the cytotoxicity of a chemical in the presence and absence of exposure to a noncytotoxic dose of simulated solar light. Cytotoxicity is measured as the inhibition of the capacity of the cell culture to take up a vital dye, neutral red, 1 day after treatment according to a published protocol [35]. This assay is considered the most appropriate in vitro screen for soluble compounds. Although the formal European Centre for the Validation of Alternative Methods validation exercise conducted on this assay indicated a sensitivity of 93% and a specificity of 84%, experience within the pharmaceutical industry suggests a much lower specificity. In the United States, for products applied dermally, a dedicated clinical trial for phototoxicity on the to-be-marketed formulation may be warranted in support of product approval.

One critical consideration in using an in vitro assay is ensuring that the effect is not cytotoxic and is phototoxic only. Because the ultraviolet A (UVA) sensitivity of cells may increase with increasing passages, cells from the lowest available passage level should be used. Furthermore, as the sensitivity of 3T3 NRU-PT is high, a positive result in 3T3 NRU-PT should not be regarded as indicative of a likely clinical phototoxic risk but rather as a flag for follow-up assessment. UVB-induced phototoxicity is rarely a problem for pharmaceuticals with systemic exposure and is more relevant for topical products. If an animal or clinical phototoxicity study has already been conducted, there is no reason to subsequently conduct either a chemical photoreactivity or an in vitro phototoxicity assay.

Hepatotoxicity

Because the liver plays a central role in the metabolism of foreign substances, the liver is one of the most important target organs for toxicity. Great importance is attached to tests using isolated liver cells. Hepatocytes obtained at necropsy or from surgical specimens via perfusion are used as primary cultures for in vitro hepatotoxicity assays. However, these cultures lose their full functionality over time, with a particularly sharp decline in their cytochrome P450 enzyme activity. Newer culture methods such as sandwich cultures or a sandwich technique involving coculture with nonparenchymatous liver cells significantly prolong the period for which metabolic activity can be maintained and can be used for in vitro hepatotoxicity assessment. Early signs of cytotoxicity involve elevation of the levels of alanine aminotransferase, aspartate aminotransferase, and lactate dehydrogenase.

Safety Pharmacology

Safety pharmacology is the study of the potential undesirable pharmacodynamic effects of a substance in relation to the dosage within the substance's therapeutic range and above. In many cases, either the parent drug or its metabolites cause activation or inhibition of nondesirable targets, leading to various safety issues. For example, activation of the human Ether-à-go-go related gene (hERG), which encodes a protein known as Kv11.1, leads to QT prolongation. Most antiarrythmics, antipsychotics, and some antibiotics (quinolones and macrolides) cause QT prolongation [36]. To assess a compound's potential for this safety issue, binding to hERG can be assayed using competition binding with labeled astemizole [37], or a thallium flux assay [38] can be adopted as a functional in vitro assay. These assays use stable cell lines, are reproducible, and can be conducted relatively inexpensively and quickly. Other targets associated with safety issues include 5HT2B (5-hydroxytryptamine receptor 2B), α1A-, α2A-, β1-, β2-adrenergic receptors, and cannabinoid receptors type 1 (CB1) and type 2 (CB2), which have been implicated in alteration of blood pressure, heart rate, euphoria, weight loss, and emesis [39]. Transporter assays are also recommended. Binding studies are typically performed to assess the effects mediated via these promiscuous targets. Because binding of a chemical with a protein does not provide information about the mode of the effect (namely, activation, or inactivation), supplementing these binding data with functional data is always a better option. The ICH guideline S7A recommends the establishment of a concentration effect relationship using a range of concentrations [40]. Furthermore, the range of concentrations used should be selected to increase the likelihood of detecting an effect on the test system. The upper limit of this range may be influenced by the physicochemical properties of the test substance and other assay-specific factors. In the absence of an effect, the range of concentrations selected should be justified. Detailed guidelines for the hERG assay are thoroughly discussed in the ICH test guideline S7B [41].

Tissue Cross-Reactivity Assay

Tissue cross-reactivity studies can provide useful information about target distribution and potential unexpected binding. For monoclonal antibodies, assessment of tissue cross-reactivity is of paramount importance. In cases where one cannot demonstrate a pharmacologically relevant species, tissue cross-reactivity (TCR) studies can be used to guide the selection of toxicology species by comparison of tissue binding profiles in human and those animal tissues where target binding is expected.

TCR studies are designed to characterize the binding of monoclonal antibodies and related antibody-like products to antigenic determinants via immunohistochemical techniques in tissues. A TCR study with a panel of human tissues is a recommended component of the safety assessment package supporting initial clinical dosing of these products. Tissue binding per se does not indicate biological activity in vivo. In addition, binding to areas not typically accessible to the antibody in vivo (e.g., cytoplasm) is generally not relevant. When there is unexpected binding in human tissues, an evaluation of selected animal tissues can provide supplemental information regarding potential correlations or lack thereof with preclinical toxicity. TCR using a full panel of animal tissues is not recommended.

THE APPLICATION OF THE PRINCIPLES OF GLP TO IN VITRO STUDIES

The OECD principles of Good Laboratory Practices (GLPs) provide a framework of administration and quality systems according to which experiments on nonclinical health, safety, and environmental effects should be planned, performed, monitored, reported, and archived in a test facility. As per OECD document ENV/MC/CHEM(98)17, "Good Laboratory Practice (GLP) is a quality system concerned with the organisational process and the conditions under which non-clinical health and environmental safety studies are planned, performed, monitored, recorded, archived and reported" [42]. Fig. 9.1 outlines the core aspects of GLP foundations laid by the OECD, which are applicable to all types of studies intended for testing of chemicals. The principles of GLP were initially developed to address the need for independent monitoring and control of experiments (more specifically, experiments involving the use of animals for toxicity studies) to generate data of a certain acceptable quality standard for regulatory agencies. In the 1970s and 1980s, several cases associated with bad practices in the industry led the US FDA and EPA (United States Environmental Protection Agency) to conclude that the toxicity data generated by some laboratories were not sufficiently reliable. This led to the elucidation of legitimate guidelines by federal agencies to enforce industry best practices for conducting and reporting toxicity studies for seeking approval of pharmaceutical agents, and these guidelines were later extended to other classes of chemicals. As a result of its strong impact and a similar turn of events within the EPA, the OECD adopted and published the principles of Good Laboratory Practices in 1997, followed by a multilateral agreement between the member countries for "Mutual Acceptance of Data". According to this legal arrangement, the data generated in a GLP-compliant facility in an OECD member nation would be accepted elsewhere in any other OECD member or nonmember nation for the purpose of scientific and regulatory consideration.

FIGURE 9.1 Concept of Organisation for Economic Co-operation and Development good laboratory practices (GLPs).

Until the 2000s, the OECD principles of GLP focused broadly on animal toxicity studies. However, simultaneously, the application of in vitro testing for hazard identification was gaining importance. Although GLP principles can be applied to in vitro studies, such applications were not under the regulatory sphere until 2004, when OECD published guidelines on "The Application of the Principles of GLP to in vitro studies" ENV/JM/MONO(2004)26. Introduction of this guidance document addresses most of the GLP aspects that require critical attention from the perspectives of facility, test materials, study execution, and quality systems [43].

The areas of critical importance from a GLP perspective may include but are not limited to the procedures, measures, and documentation for the following:

1. Monitoring of batches of components of cell and tissue culture media that are critical to the performance of the test system (e.g., bovine serum) and other materials with respect to their influence on test system performance.
2. Assessing and ensuring the functional and/or morphological status (and integrity) of cells, tissues, and other indicator materials.
3. Monitoring potential contamination by foreign cells, mycoplasma, and other pathogens or other adventitious agents as appropriate.
4. Cleaning and decontamination of facilities and equipment and minimizing sources of contamination of test items and test systems.
5. Ensuring that specialized equipment is properly used and maintained.
6. Ensuring proper cryopreservation and reconstitution of cells and tissues.
7. Ensuring proper conditions for the retrieval of materials from frozen storage.
8. Ensuring the sterility of materials and supplies used for cell and tissue cultures.
9. Maintaining adequate separation between different studies and test systems.

VALIDATION OF IN VITRO METHODS

According to the OECD document *ENV/JM/MONO(2005)14*, validation is defined as the process by which the reliability and relevance of a particular approach, method, process or assessment is established for a defined purpose [44]. This definition of validation indicates, albeit subtly, the two dimensions of an ideal validation exercise:

● Validation of the test principle
● Validation of the test method

Addressing both these approaches would constitute a comprehensive validation package for an in vitro test or any type of test. In its core purpose, validation seeks to eliminate all except true positives from the test results, and therefore the performance and extent of the validation of the test method under application have significant weight. For certain types of tests, regulators occasionally accept the results without a complete validation package, but the extent of validation required is essentially established on a discretionary basis depending on the assay, its complexity, significance and regulatory requirements. Retrospective validation may also be necessary. However, a test method is never valid in an absolute sense but only in relation to a defined purpose.

Fig. 9.2 describes the workflow, critical aspects, and process breakdown of a typical validation exercise. A validation director is appointed to execute a validation study. Based on prevalidation experiments, the scope of the assay is determined, and a validation protocol is generated that describes in detail the full course of the

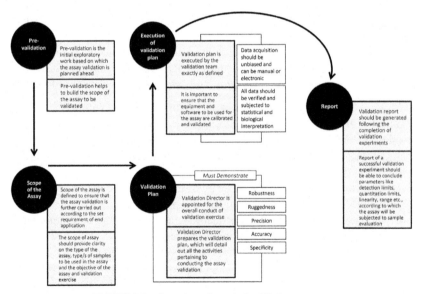

FIGURE 9.2 Overview of a typical validation process.

planned experiments to assess the validation criteria for the assay under development. The protocol is a key component of any validation experiment because it specifies the scope of the assay, objective, details of methods and subexperiments, chemicals/reagents to be used, assay conditions, equipment, software, study personnel, standard facility procedures, data recording formats, extent of quality assurance oversight (if needed), references, etc. The validation is executed exactly according to the validation protocol, and the generated data are diligently captured in the specified format. The generated data are further scrutinized to derive meaningful inferences, which are reflected conclusively in the validation report. The validation report states whether the validation exercise has been successful and also describes clearly the validated method along with the validation criteria considered (for instance, the limit of detection, working conditions, and limit of quantitation).

Validation criteria can differ from assay to assay; however, any assay should fulfill certain fundamental criteria, some of which are defined below. The definitions may suggest overlap as the differences are subtle in nature, yet the concepts are categorically standalone. The following definitions are cited from the *ENV/JM/MONO(2005)14* document prescribed by OECD [44]. There may be verbatim differences between the definitions of these concepts provided by various regulators, but the essence remains principally unchanged. For any OECD test guidelines, these definitions are invariably harmonized.

Sensitivity: The proportion of all positive/active substances that are correctly classified by the test. It is a measure of the accuracy of a test method that produces categorical results and is an important consideration in assessing the relevance of a test method.

Specificity: The proportion of all negative/inactive substances that are correctly classified by the test. It is a measure of the accuracy of a test method that produces categorical results and is an important consideration in assessing the relevance of a test method.

Robust(ness): The insensitivity of test results to departures from the specified test conditions when conducted in different laboratories or over a range of conditions under which the test method might normally be used. If a test is not robust, it will be difficult to use in a reproducible manner within and between laboratories.

Reproducibility: The agreement among results obtained by testing the same substance using the same test protocol.

Repeatability: The agreement among test results obtained within a single laboratory when the procedure is performed on the same substance under identical conditions.

Reliability: Measures of the extent that a test method can be performed reproducibly within and between laboratories over time when performed using the same protocol. It is assessed by calculating the intra- and interlaboratory reproducibility and intralaboratory repeatability.

Relevance: Description of the relationship of the test to the effect of interest and whether it is meaningful and useful for a particular purpose. It is the extent to which the test correctly measures or predicts the biological effect of interest. Relevance incorporates a consideration of the accuracy (concordance) of the test method.

Interlaboratory reproducibility: A measure of the extent to which different qualified laboratories, using the same protocol and testing the same substances, can produce qualitatively and quantitatively similar results. Interlaboratory reproducibility is determined during the prevalidation and validation processes and indicates the extent to which a test can be successfully transferred between laboratories, also referred to as between-laboratory reproducibility.

Intralaboratory repeatability: The closeness of agreement between test results obtained within a single laboratory when the procedure is performed on the same substance under identical conditions within a given time period.

Intralaboratory reproducibility: A determination of the extent that qualified people within the same laboratory can successfully replicate results using a specific protocol at different times, also referred to as within-laboratory reproducibility.

Accuracy: The closeness of agreement between test method results and accepted reference values. It is a measure of test method performance and one aspect of relevance. The term is often used interchangeably with "concordance" to indicate the proportion of correct outcomes of a test method.

REGULATORY ACCEPTANCE OF ALTERNATIVE METHODS TO ANIMAL TESTING

The objective of in vitro or alternative methods is not only to replace animal testing but also to produce predictions that are more relevant to the actual physiological process in the target species, usually humans. Thus predictors of human effects should be validated against the human effects rather than in a laboratory species or other surrogate. Several organizations internationally are proactively engaged in validating various in vitro tests as an alternative to animal testing and better association of the test results with effects in humans. Many of these organizations work closely with various national and international regulatory bodies and have been instrumental in the acceptance of a number of in vitro tests by regulatory authorities concerned with the approval of chemicals for use by humans or accidental exposure or environmental hazards.

Table 9.1 describes the list of various in vitro tests that have undergone validation and subsequent regulatory acceptance by relevant US agencies as well as the OECD Test Guideline status for use in hazard identification.

TRANSLATIONAL ASPECTS OF IN VITRO ASSAYS

Pitfalls in In Vitro Toxicology

One of the merits of in vitro systems is their simplicity. Ironically, this simplicity is also a major limitation. It is easy to understand that a simple purified enzyme or a cellular system cannot represent a complex, whole-body system. Thus the

TABLE 9.1 List of Tests Accepted by US Regulatory Agencies and Adoption by OECD as Test Guideline

Test Group	Test Method	Regulatory Acceptance US	OECD Adoption
Acute dermal systemic toxicity	In vitro dermal absorption methods	Accepted by US via OECD TG 428	OECD TG 428 (2004)
Dermal corrosivity and irritation	Corrositex in vitro membrane barrier skin corrosivity test	Accepted by US agencies in 1999; 49 CFR 173.137 (2011)	OECD TG 435 (2006, updated 2015)
	EpiSkin, EpiDerm, SkinEthic in vitro human skin model skin corrosivity test	Accepted by US via OECD TG 431; 49 CFR 173.137 (2011)	OECD TG 431 (2004, updated 2013, 2014, 2015, 2016)
	Rat TER in vitro skin corrosivity test	Accepted by US via OECD TG 430	OECD TG 430 (2004, updated 2013, 2015)
	Reconstructed human epidermis in vitro test method for skin corrosivity testing	Accepted by US via OECD TG 431 (meets performance standards 2009)	OECD TG 431 (2004, updated 2013, 2014, 2015, 2016)
	EpiSkin, EpiDerm, SkinEthic in vitro human skin model skin irritation test	Accepted by US via OECD TG 439	OECD TG 439 (2010, updated 2013, 2015)
Dermal phototoxicity	3T3 NRU phototoxicity test for skin photoirritation and application to UV filter chemicals	Accepted by US via OECD TG 432	OECD TG 432 (2004)
Endocrine disruptors	Stably transfected transactivation in vitro assay to detect estrogen receptor agonists and antagonists	Accepted by US via OECD TG 455; EPA OPPTS 890.1300 (2009)	OECD TG 455 (2009, updated 2012, 2015, 2016)
	In vitro H295R steroidogenesis assay	Accepted by US agencies via OECD TG 456	OECD TG 456 (2011)

Continued

TABLE 9.1 List of Tests Accepted by US Regulatory Agencies and Adoption by OECD as Test Guideline—cont'd

Test Group	Test Method	Regulatory Acceptance	
		US	OECD Adoption
	In vitro BG1Luc ER TA agonist and antagonist assay to identify substances that induce or inhibit human ER activity	Accepted by US agencies in 2012	OECD TG 457 (2011)
	Performance-based test guideline for human recombinant estrogen receptor binding assays	Accepted by US via OECD TG 493	OECD TG 493 (2015)
	Integrated testing strategy to identify chemicals with the potential to interact with the estrogen receptor	Accepted by EPA in 2015 as an alternative to three Tier 1 tests used in the Endocrine Disruptor Screening Program	NA
	Stably transfected human androgen receptor transactivation assay for detection of androgenic agonist and antagonist activity of chemicals	Accepted by US via OECD TG 458	OECD TG 458 (2016)
Genetic toxicity	Bacterial reverse mutation test	Accepted by US via OECD TG 471	OECD TG 471 (1997)
	In vitro mammalian chromosomal aberration test	Accepted by U.S. via OECD TG 473	OECD TG 473 (1983, updated 1997, 2014, 2016)
	In vitro mammalian cell micronucleus test	Accepted by U.S. via OECD TG 487	OECD TG 487 (2010, updated 2014, 2016)
	In vitro mammalian cell gene mutation tests using the thymidine kinase assay	Accepted by US via OECD TG 490	OECD TG 490 (2015, updated 2016)

TABLE 9.1 List of Tests Accepted by US Regulatory Agencies and Adoption by OECD as Test Guideline—cont'd

		Regulatory Acceptance	
Test Group	Test Method	US	OECD Adoption
	In vitro mammalian cell gene mutation tests using the HPRT and XPRT genes	Accepted by U.S. via OECD TG 476	OECD TG 476 (2016)
Immunotoxicity	In vitro skin sensitization test (ARE-Nrf2 luciferase test)	Accepted by US via OECD TG 442D	OECD TG 442D (2015)
	In vitro skin sensitization test (human cell line activation test)	Accepted by US via OECD TG 442E	OECD TG 442E (2016)
Ocular corrosivity and irritation	Bovine corneal opacity and permeability in vitro test method to identify severe eye irritants/corrosives or chemicals not requiring eye hazard classification	Accepted by US agencies in 2008; 2013 update accepted via OECD TG 437	OECD TG 437 (2009, updated 2013)
	Isolated chicken eye in vitro test method to identify severe eye irritants/corrosives or chemicals not requiring eye hazard classification	Accepted by US agencies in 2008; 2013 update accepted via OECD TG 438	OECD TG 438 (2009, updated 2013)
	In vitro fluorescein leakage test method for identifying ocular corrosives and severe irritants	Accepted by US agencies via OECD TG 460	OECD TG 460 (2012)
	Short-time exposure test for identification of ocular corrosives and substances not requiring ocular hazard labeling	Accepted by US agencies via OECD TG 491	OECD TG 491 (2015)

Continued

TABLE 9.1 List of Tests Accepted by US Regulatory Agencies and Adoption by OECD as Test Guideline—cont'd

Test Group	Test Method	Regulatory Acceptance	
		US	OECD Adoption
	Reconstructed human cornea-like epithelium test for identification of substances not requiring ocular hazard labeling	Accepted by US agencies via OECD TG 492	OECD TG 492 (2015)

ER, estrogen receptor; *HPRT*, hypoxanthine phosphoribosyl transferase; *NRU*, neutral red uptake; *OECD*, Organisation for Economic Co-operation and Development; *TA*, transcriptional activation; *TER*, transcutaneous electrical resistance; *TG*, test guideline; *XPRT*, xanthine phosphoribosyltransferase.
National Toxicology Program. US Department of Health and Human Services. Alternative Methods Accepted by US Agencies; September 24, 2016. http://ntp.niehs.nih.gov/pubhealth/evalatm/iccvam/acceptance-of-alternative-methods/index.html.

predicted human toxicity based on in vitro experiments may not corroborate with actual findings. Some of these limitations are discussed in this section.

1. The structure of the target (receptor, enzyme, etc.) in purified form may be different from that in the whole-animal system. This may distort the binding sites and thus the efficacy of the test item. In this situation, the in vitro toxic response may be an artifact.
2. The biokinetics of a compound may be completely different in vivo and in vitro.
3. For in vitro assays, compounds are usually dissolved in nonpolar solvents such as dimethyl sulfoxide (DMSO), which cannot be used in animal experiments. Depending on differences in formulation, the solubility of a compound (and thus absorption) may be different in vitro and in vivo.
4. The first-pass effect is not represented in an in vitro system.
5. The distribution, elimination, and metabolism of a chemical in the whole animal is not replicated in an in vitro system.

The basic challenge with any in vitro technique is interpretation of the data in relationship to the animal from which the system was derived and then, if needed, extrapolation to humans. The absence of interaction between organs or tissues and the static nature of many in vitro systems means that some observed effects are absent in life. For example, in vitro systems are isolated systems and do not include other physiological components such as transporters and blood circulation. Thus the concentrations tested may not be relevant to those expected in real life, e.g., in toxicity studies or clinically. In addition, the transient nature of incubations in most in vitro systems prevents chronic administration and the detection of progressive effects. These test systems are often technically demanding and difficult to reproduce from one laboratory to another. As noted above, the choice of time point can be critical in achieving data that can be analyzed to best advantage. In the absence of good validation work, the significance of observed differences may be misinterpreted, and erroneous conclusions may be drawn. However, the use of in vitro systems have greatly reduced the dependence on intact animal experimentation and facilitated the identification of potential concerns for hazardous molecules at an earlier stage of development, thus reducing wasted resources and animal and human exposure.

ACKNOWLEDGMENTS

The authors thank the management of Cadila Healthcare Ltd. for supporting in vitro research. Mr. Siddharth Brahmbhatt, Zydus Research Centre, aided the preparation of this chapter.

REFERENCES

[1] Russell WM, Burch RL. In: Russell WMS, Burch RL, editors. The principles of humane experimental technique. London: Methuen & Co. Ltd.; 1959.

[2] ISO 10993-1:2009 Biological evaluation of medical devices – Part 1: evaluation and testing within a risk management process. 2009. ISO 10993-1:2009 21.

[3] Use of international standard ISO-10993, 'biological evaluation of medical devices Part 1: evaluation and testing'. Center for Devices and Radiological Health, US FDA; 1995.

[4] ISO 10993-5:2009 Biological evaluation of medical devices – Part 5: tests for in vitro cytotoxicity. 2009. ISO 10993-5:2009 34.

[5] Clemedson C, et al. MEIC evaluation of acute systemic toxicity. Part VII. Prediction of human toxicity by results from testing of the first 30 reference chemicals with 27 further in-vitro assays. ATLA 2000;28:159–200.

[6] Donato MT, Jover R, Gómez-Lechón MJ. Hepatic cell lines for drug hepatotoxicity testing: limitations and strategies to upgrade their metabolic competence by gene engineering. Curr Drug Metab 2013;14:946–68.

[7] Bowtell DD. Options available–from start to finish–for obtaining expression data by microarray. Nat Genet 1999;21:25–32.

[8] Test No. 476: in vitro mammalian cell gene mutation test. OECD Publishing; 1997. http://dx.doi.org/10.1787/9789264071322-en.

[9] Test No. 479: genetic toxicology: in vitro sister chromatid exchange assay in mammalian cells. OECD Publishing; 1986. http://dx.doi.org/10.1787/9789264071384-en.

[10] Test No. 480: genetic toxicology: *Saccharomyces cerevisiae*, gene mutation assay. OECD Publishing; 1986. http://dx.doi.org/10.1787/9789264071407-en.

[11] Test No. 481: genetic toxicology: *Saacharomyces cerevisiae*, miotic recombination assay. OECD Publishing; 1986. http://dx.doi.org/10.1787/9789264071421-en.

[12] Test No. 482: genetic toxicology: DNA damage and repair, unscheduled DNA synthesis in mammalian cells in vitro. OECD Publishing; 1986. http://dx.doi.org/10.1787/9789264071445-en.

[13] Test No. 490: in vitro mammalian cell gene mutation tests using the thymidine kinase gene. OECD Publishing; 2015. http://dx.doi.org/10.1787/9789264242241-en.

[14] Test No. 474: mammalian erythrocyte micronucleus test. OECD Publishing; 2014. http://dx.doi.org/10.1787/9789264224292-en.

[15] Test No. 487: in vitro mammalian cell micronucleus test. OECD Publishing; 2014. http://dx.doi.org/10.1787/9789264224438-en.

[16] Irac. Alcohol drinking. World Health Organization; 1988.

[17] de Fouw J, International Program on Chemical Safety, United Nations Environment Programme, International Labour Organisation & World Health Organization. Chloroform. World Health Organization; 1994.

[18] Reuber MD. Carcinogenicity of captan. J Environ Pathol Toxicol Oncol 1989;9:127–43.

[19] Shi X, Castranova V, Halliwell B, Vallyathan V. Reactive oxygen species and silica-induced carcinogenesis. J Toxicol Environ Health B Crit Rev 1998;1:181–97.

[20] Ho S-M, Tang W-Y, Belmonte de Frausto J, Prins GS. Developmental exposure to estradiol and bisphenol A increases susceptibility to prostate carcinogenesis and epigenetically regulates phosphodiesterase type 4 variant 4. Cancer Res 2006;66:5624–32.

[21] Dolinoy DC, Huang D, Jirtle RL. Maternal nutrient supplementation counteracts bisphenol A-induced DNA hypomethylation in early development. Proc Natl Acad Sci USA 2007;104:13056–61.

[22] Vasseur P, Lasne C. OECD detailed review paper (DRP) number 31 on "cell transformation assays for detection of chemical carcinogens": main results and conclusions. Mutat Res 2012;744:8–11.

[23] Pratt RM, Grove RI, Willis WD. Prescreening for environmental teratogens using cultured mesenchymal cells from the human embryonic palate. Teratog Carcinog Mutagen 1982;2:313–8.

[24] Braun AG, Nichinson BB, Horowicz PB. Inhibition of tumor cell attachment to concanavalin A-coated surfaces as an assay for teratogenic agents: approaches to validation. Teratog Carcinog Mutagen 1982;2:343–54.

[25] Mummery CL, van den Brink CE, van der Saag PT, de Laat SW. A short-term screening test for teratogens using differentiating neuroblastoma cells in vitro. Teratology 1984;29:271–9.

[26] Steele VE, et al. Evaluation of two in vitro assays to screen for potential developmental toxicants. Fundam Appl Toxicol 1988;11:673–84.

[27] Festag M, Viertel B, Steinberg P, Sehner C. An in vitro embryotoxicity assay based on the disturbance of the differentiation of murine embryonic stem cells into endothelial cells. II. Testing of compounds. Toxicol In Vitro 2007;21:1631–40.

[28] Takahashi K, Yamanaka S. Induction of pluripotent stem cells from mouse embryonic and adult fibroblast cultures by defined factors. Cell 2006;126:663–76.

[29] Gurdon JB. The developmental capacity of nuclei taken from intestinal epithelium cells of feeding tadpoles. J Embryol Exp Morphol 1962;10:622–40.

[30] Daston GP, Baines D, Yonker JE. Chick embryo neural retina cell culture as a screen for developmental toxicity. Toxicol Appl Pharmacol 1991;109:352–66.

[31] Umansky R. The effect of cell population density on the developmental fate of reaggregating mouse limb bud mesenchyme. Dev Biol 1966;13:31–56.

[32] Eisenbrand G, et al. Methods of in vitro toxicology. Food Chem Toxicol 2002;40:193–236.

[33] Spielmann H, et al. The practical application of three validated in vitro embryotoxicity tests. The report and recommendations of an ECVAM/ZEBET workshop (ECVAM workshop 57). Altern Lab Anim 2006;34:527–38.

[34] Test No. 432: in vitro 3T3 NRU phototoxicity test. OECD Publishing; 2004. http://dx.doi.org/10.1787/9789264071162-en.

[35] Borenfreund E, Puerner JA. Toxicity determined in vitro by morphological alterations and neutral red absorption. Toxicol Lett 1985;24:119–24.

[36] Sanguinetti MC, Tristani-Firouzi M. hERG potassium channels and cardiac arrhythmia. Nature 2006;440:463–9.

[37] Huang S-M, Woodcock J. Transporters in drug development: advancing on the critical path. Nat Rev Drug Discov 2010;9:175–6.

[38] Titus SA, et al. A new homogeneous high-throughput screening assay for profiling compound activity on the human ether-a-go-go-related gene channel. Anal Biochem 2009;394:30–8.

[39] Bowes J, et al. Reducing safety-related drug attrition: the use of in vitro pharmacological profiling. Nat Rev Drug Discov 2012;11:909–22.

[40] European Medicines Agency. ICH S7A safety pharmacology studies for human pharmaceuticals. ICH guidelines. 2001. p. 1–11.

[41] Food, Drug Administration, HHS. International conference on harmonisation; guidance on S7B nonclinical evaluation of the potential for delayed ventricular repolarization (QT interval prolongation) by human pharmaceuticals; availability. Notice. Fed Regist 2005;70:61133–61134..

[42] OECD, 1998. OECD principles of good laboratory practice, (as revised in 1997). ENV/MC/CHEM(98)17. 1998.

[43] OECD, 2004. Advisory document of the working group on good laboratory practice: the application of the principles of GLP to in vitro studies. 2004.

[44] OECD, 2005. Guidance document on the validation and international acceptance of new or updated test methods for hazard assessment. 2005.

Chapter 10

Safety Concerns Using Cell-Based In Vitro Methods for Toxicity Assessment

Vyas Shingatgeri
Vanta Bioscience Limited, Chennai, India

Those who work the safest way, live to work another day

INTRODUCTION

Cell cultures derived from various animal/human tissues have been currently considered the best choice in the biological research and evaluation of toxicity. In vitro test systems employed to evaluate changes in biological processes using cells, cell lines, or cellular components [1] play an important role in establishing in vitro methodologies for the diagnosis, prophylaxis, and treatment of some of the dreaded diseases. In vitro toxicological screening assays not only facilitate a better understanding of cellular and functional aspects of human health and disease but also aid to discover better therapeutic interventions, reducing the use of laboratory animals in experimentation, simultaneously fulfilling the 3 "R's" (replacement, reduction, and refinement) of humane treatment of animals.

In vitro studies are often performed during drug development as part of the effort to understand the often complex disease processes and/or to understand the mechanism of actions of the drugs at the cellular and molecular levels. Biotechnology-driven large-scale cell cultures have been developed to produce more specific, less risky vaccines, lifesaving (e.g., tissue plasminogen activator) and life sustaining biosimilars or follow-on-biologics (e.g., rh-insulin, rh-erythropoietin). Cell cultures also provide important substrates for various bioassays viz., vaccine potency, vaccine efficacy, and detection of adventitious agents in the finished products. In addition, in vitro tests using microorganisms and animal/human cell lines have achieved widespread recognition in the identification of genotoxic potential of chemicals, for studying specific mechanisms and for screening purposes. More recently some significant advances have also been made to predict acute local toxic effects of xenobiotics using in vitro assays. In vitro studies for toxicity screening

In Vitro Toxicology. http://dx.doi.org/10.1016/B978-0-12-804667-8.00010-9

involve the use of variety of cells from liver, kidney, lung, heart, and brain from various species [2].

The upcoming regulations and at the same time the effort to overcome the hurdles in animal testing due to growing public concerns about animal welfare have prompted to follow the principles of 3 Rs in the use of animals in research and regulatory toxicology [3]. Recently, a draft guideline (EMEA/CHMP/CVMP/JEG-3Rs/450091/2012) has been issued on regulatory acceptance of 3 Rs testing approaches [4] and the Directive 2010/63/EU [5] on protection of animals used for scientific purposes fully endorses these principles. For regulatory acceptance, European Centre for the Validation of Alternative Methods brought together experts in the field of cell culture technology and good laboratory practice to stimulate the use and acceptance of in vitro toxicology data during the human risk assessment process. Above all, the recent draft amendments initiated during August 2016 by the United States Food and Drug Administration (US FDA) on the principles of good laboratory practice [6] (21 Code of Federal Regulations [CFR] Part 16 and 58) delineates more into Animal Welfare aspects in its legal framework emphasizing equal importance on par with quality and data integrity. Many in vitro methods have been established for safety assessment with regulatory acceptances.

Although there is an increasing need for alternative in vitro safety testing methods using cell culture, the foremost factor that hinders the progress is the integrity of a cell line with its useful characteristics that is sustained only when the culture remains free of contaminating microorganisms (pure) and other cells (authentic) and the challenge of showing stability of its characteristics on "passage" in vitro. The greatest deterrent in handling cells is their ability to harbor and provide an adequate media to multiply pathogenic organisms. Contamination with pathogenic microorganisms or cross-contamination involving other cell types can happen inadvertently while working with cells, which might go unnoticed. Cell cultures, although may appear apparently nonhazardous, can also cause potential hazards to laboratory workers. Therefore, measures should be taken to maintain the originality of the cell line and at the same time ensure safety of personnel and environment. In addition, risk may also arise from handling of cytotoxic chemicals, quickly penetrating solvents, such as use of liquid nitrogen and CO_2.

BIOSAFETY AND RISK MANAGEMENT IN IN VITRO LABORATORIES

Risk assessment is a systematic approach to identify hazards, evaluate risk, and incorporate appropriate measures to manage and mitigate risk for any work process or activity. In in vitro cell-based procedures, the process of risk assessment involves evaluation of potential exposure to risk arising from handling of cell cultures, microorganisms, and hazardous reagents/chemicals/

solvents and to prevent lab-acquired infections from bacteria, viruses, fungi, mycoplasma, human blood, unfixed tissues (primary or secondary) human cell lines, and recombinant DNA. Facility and infrastructure availability such as provision of biosafety cabinets and procedural controls like managing the use of different cell lines and culture processes at strategic time points, maintenance of biosafety cabinets, adherence to good laboratory practices, and use of process-specific additional personal protective equipment play an important role to prevent or minimize such hazards. Administrative controls including organizational policies in line with national or global guidelines also play equally important role.

Risk assessment has been identified to have a number of stages but the fundamental features of appropriate and effective risk assessment include prior planning and process of documentation after performing risk assessment, thorough communication of risk assessment to staff and maintenance, and timely review in order to avoid future occurrences [7].

Risk-Classification Based on Type of Cells

Risk	Type of Cells
High	• Uncharacterized cell line
	• Primary cells from blood, lymphoid cells and neural tissue of human or simian origin
	• Blood of human or nonhuman origin
	• Pituitary from human, caprine, and ovine origin
	• Central nervous system cells from human and bovine
	• Pathogen deliberately introduced or known endogenous contaminant
Medium	• Mammalian nonhematogenous cells such as fibroblasts and epithelial cells
	• Cell lines/strains not fully authenticated or characterized
Low	• Cells derived from avian and invertebrate tissues
	• Well characterized/authenticated finite cell lines of human or primate origin
	• Nonhuman, nonprimate cell lines which have been authenticated, have low risk of endogenous pathogens and present no human health risk

[8] Sheeley H. Risk assessment. In: Stacey GN, Doyle A, Hambleton PJ, editors. Safety in cell and tissue culture. Dordrecht: Kluwer Academics; (1998). p. 173–188.

RISK ASSOCIATED WITH CELL CULTURE

Control of Substances Hazardous to Health Regulations includes cell culture in the definition of a biological agent as they may be contaminated with adventitious biological agents such as mycoplasmas (e.g., *Mycoplasma pneumonia*) or viruses intentionally or unintentionally. The potential hazards associated with cell cultures can be related to the species and tissue of origin [9].

WHO MAY BE AT RISK?

Risks arising from mammalian cells in culture can generally be classified into three categories:

1. Risk for operator during cell cultivation
2. Risks for the recipient of the product of the cells
3. Risk for the environment in general, including animals and plants

FACTORS CONTRIBUTING TOWARD RISK IN THE CELL CULTURE

Tissue/Type, Origin, and Source of Cells

In vitro test systems could be primary cells or secondary cells. Primary cell cultures are created by growing cells directly from biopsies and often not possible to determine whether the cells are free from any infections, such as viruses. Therefore, primary cell cultures should always be handled with caution as they are a potential source of pathogens and should be considered a risk of contamination.

Established cell lines are cultures that have been grown for longer periods of time and immortalized so that they divide infinitely. In most cases, it is known whether or not they are free of contaminating biological agents. Mammalian and insect cells have very stringent requirement for growth and therefore are very susceptible to dehydration and exposure to ultraviolet radiation, outside the animals from which they are derived. So, growth and survival requirements can only be met using specialized media, the optimum temperature range, pH, and an adequate oxygen concentration. Hence, cell lines pose minimal risk to both human health and the environment. Moreover, due to immune rejection of nonself-tissue, it is highly improbable that accidental exposure would result in survival and replication in normal healthy individuals (with the exception of some tumor cells). Therefore, workers should not conduct work on their own cells and use of cells derived from other laboratory workers should be strictly avoided. Laboratory workers should never handle autologous cells or tissues [10].

Established cell lines should also not be considered as low risk unless fully authenticated because some of the established cell lines have been shown to be persistently infected (e.g., B95-8 with EBV [Epstein Barr virus], human cell line MT4 with HTLV [human T lymphocyte virus]). In particular, the continuous culture of permissive (CD4+) primate cell lines must be carefully assessed and the possibility of contamination with an inadvertent coculture of human immunodeficiency virus (HIV) or simian immunodeficiency virus (SIV) should be taken into account. If HIV and SIV are known to be present then a minimum of containment level three will be required.

The extent of risk can be depicted in decreasing order: primary cell cultures>continuous cell lines (immortalized cells)>intensively characterized cells (including human diploid fibroblasts). Commercial suppliers often provide biocontainment recommendations based on characterization of the cells. The recommendation is a good starting point, but may need to be modified or critically reviewed based on other risk assessment considerations in the utility environment. When dealing with primary human cells, risk assessments should also include the quantity of cells per specimen, the number of specimens from different individuals, and the level of risk represented by the population from which specimens are obtained. The extent of risk can be depicted in hematogenous cells and tissue (blood, lymphoid tissue), neural tissue, endothelium, gut mucosa, epithelial cells, and fibroblasts.

VIRAL CONTAMINANATION

1. Main viral contaminants in human tissue:
   Viruses pathogenic to humans are one of the most likely biohazards presented by cell cultures.
   a. Hepatitis viruses
   HBV (hepatitis B virus), HCV (hepatitis C virus), HDV (hepatitis D virus), HEV (hepatitis E virus)
   b. Human retroviruses
   HIV-1, HIV-2, HTLV-1, HTLV-2, SIV
   c. Herpes viruses
   EBV, CMV (cytomegalus virus), human herpes virus-6, human simian virus 1 and 2
   d. Popovirues
   Different human papillomavirus (HPV)
   e. Prions
   Infectious prion proteins
2. Viral contaminants in primate tissues causing human disease
   a. RNA viruses:
   Flaviviruses, filoviruses, simian hemorrhagic virus, rabies virus, HAV, poliovirus
   DNA-Viruses: Herpes viruses (herpes-b-virus and others), *Mossuscum contagiosum*
   b. Mice viruses pathogenic for humans
   Lymphocytic choriomeningitis virus
   Hantan virus (hemorrhagic fever with renal syndrome)
   Sendai virus (murine parainfluenza virus type 1)

The main risk associated with cell cultures is a result of their ability to sustain the survival and/or replication of a number of adventitious agents. The possibility of adventitious infection of cell lines should not be undermined. This is highlighted by several recent reports of laboratory cultured cell lines infected with murine retroviruses [11], which involve a report of contamination with gamma retrovirus xenotropic murine leukemia virus-related virus. The xenotropic retroviruses have a unique property that they can infect foreign cells, such as human cells, but do not reinfect mouse cells.

Potential laboratory hazards associated with human cells and tissues include the blood borne pathogens (viruses), HBV, HIV, HCV, HTLV, EBV, HPV, and CMV as well as agents such as *Mycobacterium tuberculosis, Mycoplasma pneumoniae* that may be present in human lung tissues. Cells immortalized with viral agents such as SV-40 (simian virus-40), EBV adenovirus, or HPV, as well as cells carrying viral genomic material also present potential hazards to laboratory workers. Tumorigenic human cells also are potential hazards as a result of self-inoculation [12] and there is a report on development of tumor from an accidental needle stick [13]. Similarly, nonhuman primate (NHP) cells blood, lymphoid, and neural tissues should always be considered potentially hazardous.

The occurrence of cross-contaminated or mislabeled cell lines continues to be one of the serious scientific issues that has so far been received insufficient attention [14,15]. The primary cell lines derived from blood or neural tissue and cell lines that have not been full authenticated or characterized are more likely to harbor adventitious agents than other cell lines.

Cells and tissues originating from nonhuman primates are more likely to contain biological agents classified as potentially hazardous to human health. Other primate cells and tissues also present risks to laboratory workers [16]. Human cells of blood origin represent a relatively high potential risk due to possibility of the presence of blood borne pathogens. However, well-characterized and quality controlled human cell lines will generally represent a lower degree of risk than primary cells. Cell/tissue source will indicate the potential contaminants and the potential for expression or reactivation of latent viruses.

The Occupational Safety and Health Administration (OSHA) standard for blood borne pathogens (BBP) may apply to laboratory workers who handle certain cell lines. Established human cell lines, which are characterized to be free of contamination from human hepatitis virus, human immunodeficiency viruses, and other recognized blood borne pathogens are not subject to BBP standard.

All primary human cell explants from tissues and subsequent in vitro passages of human tissue explant cultures (nontransformed, referred to by OSHA as human cell strains) must be regarded as containing blood borne pathogens and are subject to BBP standard, unless characterized by testing to be free of blood borne pathogens (Biosafety of Mammalian Cell Cultures: Proceedings of Basel forum on biosafety [17]).

Problems Associated With Viral Contaminations

Virus contaminated cell cultures represent a risk for the operators, collaborators, patients, as well as noncontaminated cell cultures. Viral infection can be highly pathogenic and because there are no effective treatments available, consideration should be given for prevention of viral infections. In contrast to bacterial and fungal contamination, viral contamination cannot be easily detected because they cannot be observed by normal light microscopy. It is only when a viral contamination leads to morphological modification of the cultured cells, such as cytopathic effect, that a contamination by a virus can be suspected. Silent infection by viruses with no observable morphological modifications of the infected cell is clearly of greater concern.

Cells contaminated with some viruses can show change in their susceptibility to infection by other viruses. For example, some safety testing protocols require indicator cells to be used to show the presence of virus if these are chronically infected by viruses, which reduce their susceptibility to other virus species. This in turn can lead to false negative results because the virus to be detected can no longer infect the indicator cells [18].

As a general statement, cell lines contaminated by viruses cannot be treated to become virus free. The result of this is that potentially valuable cell lines will have to be discarded and replaced by new noncontaminated cells. However, one of the few exceptions to this rule is the case of lactate dehydrogenase virus. Cultures of cells recovered a passage from infected animals may contain the virus. However, as this virus cannot infect the cells, it will be diluted during subsequent in vitro passaging and thus will be lost [19,20].

Origin of Viral Contaminations

Viruses can be introduced by a number of different routes and the knowledge of the same prompts to avoid infection. Cell cultures can become contaminated by the following means: first, they may already be contaminated as primary cultures (because the source of the cells was already infected), second, they were contaminated due to the use of contaminated raw materials, or third, they were contaminated via an animal passage. The identified routes of infection are invariably the use of already contaminated cells with the exogenous virus at source for the production of cells. This happens when, for example, the donor animal is already infected from which cells are being explanted or the preculture (in vitro) or an in vivo passage in an animal that leads to virus contamination.

Several cell lines of importance such as murine hybridomas or Chinese hamster ovary cells contain endogenous retroviruses and can produce retroviral particles during production. The cell cultures can be contaminated by viruses, which were present in the animal-derived materials used in the manipulation or for the growth of the cells. These types of materials include serum or trypsin.

MYCOPLASMAS

Major agents of concern are viruses but other agents such as mycoplasmas (*mycoplasma pneumoniae*) should also be considered. Mollicutes, often named mycoplasmas, are a group of extremely small bacteria. Some species are plant and insect pathogens and others cause veterinary and human diseases and many colonize plants, animals, or man as pathogenic commensals. Cultivation of most mycoplasmas in artificial media requires several supplements, for example, cholesterol. With the development of eukaryotic cell culture containing these supplements, mycoplasmas have been recognized as potential culture contaminants [21,22]. In contrast to direct visible contaminations with bacteria or fungi, which overgrow and kill the cells within days, mycoplasmas are able to persist in permanent cell cultures undetected for years. With the recent expansion of cell culture in biotechnology the contamination by mycoplasmas has become a new threat. In fact, as easier as the occurrence of contamination, as difficult as the detection and elimination of the same.

How Does Mycoplasma Enter Cell Cultures?

The factors responsible for the source of mycoplasma contamination are as follows:

- Spreading by handling of a previously contaminated cell lines (the so-called "kind gifts" from other laboratories are in fact often "Trojan horses")
- Contamination with human commensal mycoplasmas from laboratory personnel
- Import to laboratory by establishing a primary cell culture of an infected animals
- Use of contaminated supplements (serum, cytokines, supernatants of infected cells)
- Splitting adherent cells with contaminated trypsin

Mycoplasma-infected host cells can change several properties such as growth, metabolism, morphology, and genome structure [23]. Degradation of arginine and purines by mycoplasmas may inhibit the synthesis of histones and nucleic acid, followed by chromosome aberrations of the host cells. Some mycoplasmas produce hydrogen peroxide, which is directly toxic to the cells (Biosafety of Mammalian cell cultures Proceedings, Basel Forum on Biosafety [17]).

The nature of cell culture systems also means that they are often able to sustain and allow amplification of such agents during use. In addition, contaminated cells may also present other hazards to human health, such as ability to produce toxins or allergens. An assessment of the risks must therefore be undertaken before work involving any cell culture commences.

CULTURE MEDIA

Media used in cell line derived from an animal may also act as a potential source of contamination. Media contaminants may grow when introduced to cell culture. The most common contaminant is bovine viral diarrhea virus (BVDV).

Most media, sera, and other animal-derived biologicals are not heat sterilizable and require membrane filtration (sometimes radiation is also used) to remove biological contaminants. Products filter sterilized in the laboratory should always be tested for sterility before use (discussed in detail later); commercially produced sterile products are tested by the manufacturer before being sold. Although filtration through 0.2 μm membranes is very effective in removing most biological contaminants, it cannot guarantee the complete removal of viruses and mycoplasmas, especially in sera. In an excellent review of the rates and sources of mycoplasma contamination, Barile and coworkers reported that 104 out of 395 lots (26%) of commercial fetal bovine sera tested were contaminated by mycoplasma. Although these products are no longer a major source of mycoplasma contamination, they must still be considered as potential sources to be evaluated whenever mycoplasmas are detected in cultures [24].

TRANSMISSION INVOLVING USE OF NEEDLES AND SHARP OBJECTS DURING MANIPULATIONS

The possibilities of exposure are due to puncture or cut (needle stick, contaminated broken glass), contact with broken skin, splash to mucous membranes of eye, nose, and mouth. Due to mechanical injuries, cell material may be accidently transferred directly to an operator's tissue and/or blood stream [25].

FETAL BOVINE SERUM

Four viral contaminants have been routinely detected in unprocessed and commercial lots of fetal bovine serum, bacteriophage, infectious bovine rhinotrachetis, parainfluenza-3, and BVDV. Among them, BVDV is consistently present in the majority of commercial lots of fetal bovine serum (FBS) [26]. BVDV affects production of biological reagents and the results of diagnosis [27]. The filtration process used in the presence of commercial FBS abrogates most viral agents that may be present in raw FBS, but BVDV may pass through the filters because of the pore size. By indirect immunofluorescence techniques, in cultures maintained with high concentrations of serum for at least 2 weeks, viral antigen for BVDV detection was shown up to 80%, which showed that most of the commercial FBS contains BVDV. To eliminate the problem of BVDV contamination in FBS, gamma-irradiated FBS is used.

HANDLING CELLS OUTSIDE BIOSAFETY CABINETS

Manipulation of infected cell culture outside the biosafety cabinets due to procedural requirements poses a safety risk to the operator. Flow cytometric analysis and sorting of cell populations constitute a special case of cell manipulation in which cells are handled outside biosafety cabinets. In addition, tissues/cells used in diagnostics and pathological laboratories are used outside biosafety cabinets. The use of fixatives in many cases is not appropriate (e.g., viable cell sorting for subsequent further cell culturing) and the risk of aerosol formation can be particularly high, especially during experimentation and upon instrument failure such as clogging of nozzles.

ALTERING CULTURE CONDITIONS

Changing the availability of cell-specific nutrients, growth factors, signal molecules, or adopting coculture techniques may have significant effects on handling of animal cell cultures as it may result in an altered neoplasia [28], altered expression of proto (oncogenes), or cell surface glycoproteins and release of endogenous viruses [29]. As a consequence, changing culture conditions may lead to altered susceptibility of cultured cells to biologic agents such as viruses [30,31].

Altering culture conditions may produce surprising results. Changes in temperature, supplements, or growth surfaces can induce changes in oncogene expression or induce expressions of endogenous viruses or alter interactions between recombinant virus and endogenous genomic provirus.

RISK ASSOCIATED WITH PROCEDURES INVOLVING CRYOBANKS/LIQUID NITROGEN

When a cell line is first established, a cryopreserved archive stock, at low passage, is essential to protect against accidental loss of the culture. Preparation of cryopreserved cell banks that are well characterized is central to resolve such concerns. Liquid nitrogen presents a serious frostbite hazard to the handler and protective gloves, masks, and apron designed for cryogenic work should be readily available and used whenever working with storage and supply vessels. On occasions, ampoules stored in liquid phase can explode as a result of rapid vaporization of liquid nitrogen trapped in ampoule on warming, leading to serious penetration injuries and this is especially grave when working with infectious materials.

Cell banks stored for long-term use, including patent deposits, will require consideration of the additional important challenges, and particular attention should be paid to the maintenance and monitoring of storage conditions.

RISK ASSOCIATED WITH PROCEDURES INVOLVING CO_2 CYLINDERS

The gas atmosphere and medium composition are critical factors in the in vitro development. During the cell culture processes, CO_2 and bicarbonate salts are added from a buffer solution for the perfect maintenance of pH according to media or organism requirements. Carbon dioxide, in all forms, can be used for many purposes and it also aids in the cell metabolism. It is important to use its capabilities correctly in order to achieve the desired effect and eliminate hazards. Working safely with carbon dioxide means understanding the characteristics of this gas and taking suitable safety precautions. Carbon dioxide is colorless and essentially odorless and tasteless and it is, therefore, impossible to detect with human sense. Carbon dioxide is considered nontoxic and not a hazardous substance as defined by the Dangerous Substances Respiratory Preparatory Directives. The ambient air contains about 0.03% carbon dioxide. This concentration is essential for life because it stimulates the respiratory center and controls the volume and rate at which the air-breathing organisms respire. At higher concentrations, CO_2 can be life threatening, i.e., when the air reaches 3%–5%.

CO$_2$ will cause headache, respiratory disturbances, and discomfort. At 8%–10%, cramps, unconsciousness, respiratory arrest, and death can occur. When cryogenic liquefied CO_2 or CO_2 that has been cooled by expansion comes in contact with human skin as a spray or snow it can produce painful "cryogenic burns". Sensitive tissues such as the cornea are particularly at risk and large areas of freeze burning can cause death (Linde safety advice 1, handling of cryogenic liquefied gases). Inhalation of CO_2 in concentrated form is dangerous to life. Solid carbon dioxide or cardice commonly used for shipment of cryopreserved culture carries similar risks of frostbite and asphyxiation similar to those of liquid nitrogen and its use should be assessed with caution. Unauthorized transfer of carbon dioxide from one gas cylinder to another constitutes a safety risk for the following reasons. Cylinders being filled must meet certain requirements so that they can reliably withstand pressure. In general, only the properly trained personnel of an authorized filling facility can determine whether a cylinder is suitable for use [32] (Linde Safety advice. 12—Working with CO_2).

RISK ASSOCIATED WITH PROCEDURES INVOLVING TOXIC COMPOUNDS

Toxic compounds such as cytotoxic drugs may have carcinogenic, mutagenic, and/or teratogenic potential. With direct contact they may cause irritation of skin, eye, mucous membrane, ulceration, and necrosis of tissue. Factors such as physical form, concentration, mode of action, and frequency of use of a particular compound affect the risk posed by a particular compound. For a biological toxin, it is important to consider specific activity of the material. Toxic effect due to protein denaturation or membrane damage will often be predictable and readily quantified. In case of compounds whose biological activity is not well characterized, e.g., potential carcinogen like ethidium bromide is used as a DNA-binding dye in various in vitro assays or where the material may have long-term or more subtle effects as in case of teratogens such as selenium, the risk is more than one could decipher. Preparations that may be inhaled (volatile solutions, fine powders) or can readily be penetrating the skin will carry high risk, especially if used with substances that can facilitate penetration into epithelial surfaces (e.g., glacial acetic acid, dimethyl sulfoxide). Fumigation of a cabinet or room with formaldehyde in the presence of hypochlorite produces carcinogenic substance by the reaction between both the chemicals. Therefore, when fumigation is carried out with formaldehyde all the hypochlorite must first be removed. Aldehydes are irritants and may cause sensitization problems [33]. Phenolic-based disinfectants are not supported as part of European Union biocidal product directives' review program. High-energy physical processes such as centrifugation and homogenization will also increase the risk of exposure through generation of aerosols.

PHYSICAL HAZARDS AND OTHERS

Besides exposure to chemicals and biological agents, laboratory workers may also be exposed to a number of physical hazards. Some of the common physical hazards that they may encounter include ergonomics, ionizing radiations, nonionizing radiations, and noise hazards.

Ergonomic Hazards

Laboratory workers are at risk of repetitive motion injuries during routine laboratory procedures such as pipetting, working at microscopes, operating microtomes, using cell counters and key boarding at computer workstations. Repetitive motion injuries develop over time and occur when muscle and joints are stressed, tendons are inflamed, nerves are pinched, and the flow of blood is restricted. Standing and working in awkward positions in front of laboratory hoods/biological safety cabinets can also present ergonomic problems.

Ionizing Radiations

Ionizing radiation sources are found in a wide range of occupational settings, including laboratories. These radiation sources can pose a considerable health risk to affected workers if not properly controlled.

Nonionizing Radiations

Nonionizing radiations do not have enough energy to cause ionization in the matter through which they pass, they are electromagnetic. There are various instruments in the in-vitro lab that generate nonionizing radiation, e.g., ultraviolet lights, ultrasonic baths, microwave ovens. They generally have less potential to cause serious health effects than ionizing radiations; their health effects are mainly due to internal body heating and induced electric currents.

Noise Hazards

In the workplace, noise pollution is generally considered to be an issue once the noise level is greater than 55 dBA. Almost all equipment that is employed for the direct performance of the function of the laboratory generates some noise, e.g., fume hoods, refrigerators, freezers, ultracentrifuges, tissue homogenizers, stirrer motors. The temporal noise level of the various noise generating systems should likewise be considered. Some systems will generate a steady noise (e.g., a fume hood), others are operating on a periodic basis (e.g., a refrigerator) [34].

RECOMMENDED PRACTICES

Personnel working in in vitro laboratory should be thoroughly familiar with the possible biological hazards and their consequences. Every researcher in the laboratory should know the basic principles of biosafety by heart and act accordingly. Only by adequate training and full adherence to the biosafety procedures, a working ambience could be created that is the safest for the workers, colleagues, and laboratory environment.

Each institution should conduct a periodic risk assessment based on the origin of the cells or tissues (species and tissue type) as well as the source (recently isolated or well-characterized). A principal investigator is required to conduct a risk assessment in support of any proposed work that is subject to review and approval by the Institutional Biosafety Committee.

Assessment of potential for pathogenic agent is of utmost concern. In general, the potential for exposure to pathogenic agents and or tumorigenicity must be considered, whether arising from cells themselves or introduced through laboratory practices. Tumorigenicity must be considered but the risk could be of minimal level because there has been only one documented case of a researcher developing a tumor following an accidental needle stick. Human and other primate cells should be handled using Biosafety laboratory (BSL-2) practices and containments. All work should be performed in a Biosafety Cabinet and all materials involved should be decontaminated by autoclaving or disinfection before discarding [13,16,35–38]. BSL-2 recommendations for personnel protective equipment such as laboratory coats, gloves, and eye protection should be rigorously followed. All laboratory staff working with human cells and tissues should be enrolled in an occupational medicine program specific for blood borne pathogens and should work under the policies and guidelines established by the institution's Exposure Control Plan (ECP). Laboratory staff working with human cells and tissues should provide a baseline serum sample, be offered hepatitis B immunization, and be evaluated by a health care professional following an exposure incident or periodically. Similar programs should be considered for work with NHP blood, body fluids, and other tissues.

It is important to use cells that have a clear lineage, i.e., cells that have been grown in the laboratory. This is done by beginning with an authenticated seed stock. There are many cell repositories. Some examples are the American Type Culture Collection (ATCC) and the Cornell Institute for Medical Research in the United States. In the U – EU, European Culture for Cell Cultures, and the German Collection of Microorganisms and Cell Cultures, and in Japan, the Riken Bioresource Center. It is important to avoid borrowing cells from the lab down the hall. It has been estimated that 15%–19% of published cell culture based research uses cells that are misidentified or cross-contaminated.

Materials sourced from the originator, an established supplier or national cell culture collection such as ATCC and European Collection of Authenticated Cell Cultures (*ECACC*), will normally be supplied with extensive information regarding these hazards and should have been screened for human pathogens. Peer-reviewed publications describing the cells will also be a useful source of similar information. Referring to these documents prompts early judgments regarding risks of preventing these kinds of contamination with more confidence. Cells from other source may have a high passage number and this may not have been accurately recorded, which eventually render our judgments about risk of infection hard to make. Of course, the potential for culture conditions to amplify any contaminating agents must be preassessed and measures be drawn toward controlling this, wherever necessary and appropriate.

For genetically modified organisms, it is the responsibility of the individual cell culture user, and institution to ensure compliance with the regulations set by authorities of a particular country.

All personnel involved in flow cytometry must be aware of potential hazards associated with this predicament and only experienced and well-trained operators should perform potentially biohazardous cell sorting. General recommendations approved by international society of analytical cytology should be adhered to set a basis for biosafety guidelines in flow cytometer laboratories [39]. Basic procedures and methods have also been described in the recommendations to ensure sorting of cell material under biosafety conditions [40,41].

When a laboratory is being designed, the manager should work with the architects to ensure that the facility is designed to meet OSHA standards and provide a safe workplace. If a new building is being constructed, all of the above sources of noise should be considered.

OSHA's ionizing radiations standard [42] (29-CFR 1910.1096) sets forth the limitations to exposure to radiations from atomic particles. Any laboratory possessing or using radioactive isotopes must be licensed by the Nuclear Regulatory Commission (NRC) and/or by a state agency that has been approved by the NRC and [43,44] (10 CFR 31.11 and 10 CFR 35.12).

Standards for nonionizing radiation are set by legislation or recommended by the Radiation Protection Division of the Environment Protection Authority, the National Health and Medical Research Council (NHMRC), Australian Radiation Protection and Nuclear Safety Agency (*ARPANSA*), Codes of Practice, Australian Standards, and Worksafe Australia.

Threshold limit values and occupational exposure levels for chemical substances are set for certain compounds/gases and will provide information on the risk of personnel due to occupational exposure.

Cytotoxic drugs are categorized as regulated waste and, therefore, should be disposed according to national or international regulatory requirements. Before containers are removed from the hood or cabinet, the exterior of closed primary container should be decontaminated and placed in a clean secondary container. Toxins should be transported only in leak/spill-proof secondary containers [7].

CONTAINMENT MEASURES FOR WORKING WITH CELL CULTURES

Where adventitious agents (or gene sequences from them) may be present in cells, containment measures should be applied, which are commensurate with the risk. Where a cell line is deliberately infected with biological agents, or where it is likely that the cell line is contaminated with a particular agent, the containment level used must be appropriate for work with that agent. Following are baseline containment measures required for working with cell cultures.

Recommended baseline containment measures for work with cell cultures.

Hazard	Cell Type	Baseline Containment
Low	Well characterized or authenticated or continuous cell lines of human primate origin with a low risk of endogenous infection with a biological agent presenting no apparent harm to laboratory workers and which have been tested for the most serious pathogens.	BSL-1
Medium	Finite or continuous cell lines/strains of human or primate origin not fully characterized or authenticated, except where there is a high risk or endogenous biological agents, e.g., blood borne viruses.	BSL-2
High	Cell lines with endogenous biological agents or cells that have been deliberately infected.	Containment level appropriate to the agent. For Example, T cells infected with HIV would require Biosafety level-3
	Primary cells from blood or lymphoid cells of human or simian origin	Containment level appropriate to the risk

N.B: Any work that could give rise to infectious aerosols must be carried out in a suitable containment, e.g., a microbiological safety cabinet.

STATE AND FEDERAL REGULATIONS

Laboratory safety is governed by a numerous local, state, and federal regulations and workers have the right to a safe workplace. The occupational safety and health act of 1970 (OSH Act) was passed to prevent workers from being killed or seriously harmed at work. The law requires employers to provide their employees with working conditions that are free of known dangers.

Over the years, OSHA has promulgated rules and published guidance to make laboratories safe for the lab personnel. OSHA standards apply both to laboratories and laboratories activities (OSHA, Laboratory Safety Guidance, [45]).

After the establishment of OSHA, working injuries and illness have reduced to 50% from 6.6 million in 1995, 5.7 million in 2000 to 3.7 million in 2010, which proves the effectiveness of the OSHA's guidance and implementations.

OSHA STANDARDS

Laboratory Standards

The CFR [46] (29 CFR 1910.1450) requires that an employer designates a Chemical Hygiene Officer and have a written Chemical Hygiene Plan (CHP) and actively verify that it remains effective. The CHP must include provisions for worker training, chemical exposure monitoring where appropriate, medical consultation when exposure occurs, criteria for use of personal protective equipment and engineering controls, special precautions for particularly hazardous substances, and a requirement for a chemical hygiene officer responsible for implementation of CHP. The CHP must be tailored to reflect the specific chemical hazards present in the laboratory where it is to be used.

Hazard Communication Standards

The CFR [47] (29 CFR 1910.1200) also called as Hazcom standard requires evaluating the potential hazard of chemicals and communicating information concerning those hazards and appropriate protective measures to the employees promptly. The OSHA standard requires manufactures and importers of hazardous chemicals to provide material safety data sheets to the users of chemicals describing potential hazards and other information.

The Blood Borne Pathogens Standard

The CFR [48] (29 CFR 1910.1030), including changes mandated by the Needle Stick Safety and Prevention Act of 2001, requires employers to protect workers from infection with human blood borne pathogens in the workplace. The standard covers all workers with reasonable anticipated exposure to blood or other potentially infectious materials. It requires that information and training be provided before the employee begins the work that may involve occupational exposure to blood borne pathogens annually and before the employee is offered hepatitis B vaccination. The blood borne pathogen standard also requires advance information and training for all workers in research laboratories who handle HIV or HBV. The standard was issued as a performance standard, which means that the employer must develop a written ECP to provide a safe and healthy work environment.

The Personal Protective Equipment Standard (29 CFR 1910.132)

This requires the employer to provide and pay for Personal Protective Equipment (PPE) and ensure that it is used wherever hazards of processes or environment, chemical hazards, radiological hazards, or mechanical irritants are encountered in a manner capable of causing injury or impairment in the function of any part of the body through absorption, inhalation, or physical contact.

The Eye and Face Protection Standard (29 CFR 1910.133)

This requires the employer to ensure that each prospective worker uses appropriate eye of face protection when exposed to eye or face hazards from flying particles, molten metal, liquid chemicals, acids or caustic liquids, chemical gases, or vapors or potentially injurious light radiation.

The Respiratory Protection Standard (29 CFR 1910.134)

This standard requires that each worker to be provided with a respirator when such equipment is necessary to protect the health of such individuals. The employer must provide respirators that are appropriate and suitable for the purpose intended.

The Hand Protection Standard (29 CFR 1910.138)

This requires employers to select and ensure that workers use appropriate hand protection when their hands are exposed to hazards such as those from skin absorption of harmful substances, severe cuts or lacerations, severe abrasions, punctures, chemical burns, and harmful temperature extremes.

The Control of Hazardous Energy Standard (29 CFR 1910.47)

The Lockout/Tag out standard establishes basic requirements for locking/and or tagging out equipment during installation, maintenance, testing, repair, or construction operations. The primary purpose of the standard is to protect workers from the unexpected energization or startup of machines or equipment or release of stored energy. In addition, other OSHA standards pertain to electrical safety, fire safety and slips, trips and falls. Details of the various acts can be assessed from the OSHA laboratory Safety Guidance [45].

CONCLUSIONS

In vitro test systems involving cells, cell lines, or cellular components play an important role in diagnosis, prophylaxis, and treatment of some dreaded diseases. Similarly, in vitro toxicological screening assays facilitate a better understanding of the cellular and functional aspects of human health and disease for better therapeutic interventions by reducing the use of animals in research. Therefore, in vitro studies are most often performed during early drug development as part of the efforts to understand complex disease processes and/or mechanism of action. Although there is an increasing need for alternate in vitro methods using cells, cell lines, and cultures, the factor that hinders the progress is the integrity of a cell lines, which could sustain only when the cultures remain free from contaminations through passaging. Contamination with pathogenic microorganisms or cross-contamination involving other cell types can happen inadvertently, while working with cells, which might be unnoticed. Cell cultures, although may appear apparently nonhazardous, can also cause potential hazards to laboratory workers. Therefore, measures should be taken to maintain the originality of the cell lines and safety of personnel and environment. In addition, risk may also arise from handling of cytotoxic chemicals, quickly penetrating solvents, use of liquid nitrogen and CO_2, while working in in vitro laboratories.

As there are various possibilities with regard to exposure to biological and chemical hazards during in vitro toxicology work, the scientists and other laboratory workers are advised to strictly follow recommendations for cell culture and be cautious because any cell is a potential virus carrier. Although routine virus testing is mandatory and a valuable means for reducing the risk of using virus contaminated cell lines, it does not provide absolute safety because of possibility of emerging viruses and the permanently existing risk of contaminations by adventitious agents. Thus, the users and producers of biotech products by using animal cells have to be cautious and ensure absence of any adventitious agents/viruses by any means.

Personnel workers in in vitro laboratories should be thoroughly familiar with the biological hazards and their consequences. Cell culture training courses covering the basics of tissue culture safety should be offered to staff and also regular updates of recommendations with regard to biosafety aspects should be provided. Use of well-written standard operating procedures and well-defined risk assessment strategies also reduce the potential harm related to biohazards. Good laboratory practice and good bench techniques such as ensuring that work areas are uncluttered, reagents are correctly labeled and stored, are also important contributors for reducing risk and making the laboratory a safe work environment. Treatment of all human blood and unfixed tissues should be carried out with utmost care similar to what is expected with dealing of HIV and HBV. In addition, it is recommended that people working in laboratories where primary human material is used are appropriately vaccinated and well protected from the hazardous and pathogenic organisms. Chemical hazards are primarily chemical reagents for which the level of risk, particularly with long-term exposure, may be difficult to ascertain. Comprehensive and effective training programs, ensuring adequate attentiveness, monitoring of the procedures, and proper documentation are essential components to ensure both quality and safety of laboratory work. There should be periodic hazard mapping of a facility and also annual assessments of the same to ensure safety of the people, premises, and the environment. Every researcher working in in vitro laboratories should know the basic principles of biosafety thoroughly and act accordingly without compromising it at any cost and any point of time, after all, for the benefit of his or her own safety.

REFERENCES

[1] Krewski D, Acosta Jr D, Andersen M, Anderson H, Bailar 3rd JC, Boekelheide K, Brent R, Charnley G, Cheung VG, Green Jr S, Kelsey KT, Kerkvliet NI, Li AA, McCray L, Meyer O, Patterson RD, Pennie W, Scala RA, Solomon GM, Stephens M, Yager J, Zeise L. Toxicity testing in the 21st century: a vision and a strategy. J Toxicol Environ Health B Crit Rev 2010;13:51–138.

[2] Stacey G. Fundamental issues for cell line banks in biotechnology and regulatory affairs. CRC press LLC; 2004.

[3] De Angelis I, Gemma S, Caloni F. CELLTOX: the Italian association for in vitro toxicology. Altern Animal Exp 2010;27:59–60.

[4] Guideline on regulatory acceptance of 3R (Replacement, Reduction, Refinement) testing approaches EMA/CHMP/CVMP/JEG-3Rs/450091/2012.

[5] Directive 2010/63/EU Legislation for the protection of animals used for the scientific purposes.

[6] CFR – Code of federal regulations title 21-parts 16 and 58 good laboratory practice for non-clinical laboratory studies; proposed rule. Fed Regist Wednesday, August 24, 2016;81(164).

[7] Jones B, Stacey G. Safety considerations for in vitro toxicology testing. In: Stacey GN, et al., editor. Cell culture methods for in vitro toxicology. Kluwer academic publishers; 2001. p. 43–66.

[8] Sheeley H. Risk assessment. In: Stacey GN, Doyle A, Hambleton PJ, editors. Safety in cell and tissue culture. Dordrecht: Kluwer Academics; 1998. p. 173–88.

[9] Frommer W, Archer L, Boon B, Brunius G, Collins CH, Crooy P, Doblhoff-Dier O, Donikian R, Economides J, Frontali C. Safe biotechnology (5), recommendation for safe work with animal and human cell cultures concerning potential human pathogens. Appl Microbiol Biotechnol 1993;39:141–7.

[10] Doblhoff-Dier O, Stacey G. Cell lines: applications and biosafety. In: Fleming D, Hunt D, editors. Biological safety: principles and practices. Washington, DC: ASM Press; 2000. p. 221–39.

[11] Paprotka T, Delviks-Frankenberry KA, Cingöz O, Martinez A, Kung HJ, Tepper CG, Hu WS, Fivash Jr MJ, Coffin JM, Pathak VK. Recombinant origin of the retrovirus XMRV. Science 2011;333:97–101.

[12] Weiss RA. Why cell biologists should be aware of genetically transmitted viruses. Natl Cancer Inst Monogr 1978;48:183–9.

[13] Gugel EA, Sanders ME. Needle-stick transmission of human colonic adenocarcinoma (letter). New Engl J Med 1986;315:1487.

[14] MacLeod RA, Dirks WG, Matsuo Y, Kaufmann M, Milch H, Drexler HG. Widespread intraspecies cross-contamination of human tumor cell lines arising at source. Int J Cancer 1999;83:555–63.

[15] Stacey G, Masters JRW, Hay RJ, Drexler HG, Macleod RAF, Freshney IR. Cell contamination leads to inaccurate data: we must take action now. Nature 2000;403:356.

[16] Caputo JL. Safety procedures. In: Freshney RI, Freshney MG, editors. Culture of immortalized cells. New York: Wiley-Liss; 1996. p. 25–51.

[17] Biosafety of mammalian cell cultures: proceedings of basel forum on biosafety. Basel, Switzerland: Agency for Biosafety Research and Assessment of Technology Impacts of the Swiss Priority Programme Biotechnology; October 28, 1993.

[18] Merten OW. Virus contamination of cell cultures-a biotechnology view. Cytotechnology 2002;39:91–116.

[19] Nakai N, Kawaguchi C, Kobayashi S, Katsuta Y, Watnabe M. Detection and elimination of contaminating microorganisms in transplantable tumors and cell lines. Exp Anim 2000;49:309–13.

[20] Nicklas W, Kraft V, Meyer B. Contamination of transplantable tumors, cell lines and mono-clonal antibodies with rodent viruses. Lab Anim Sci 1993;43:296–9.

[21] Del Gludice RA, Hopps HE. In: McGarrity GJ, Murphy D, Nichols WW, editors. Mycoplasma injection of cell cultures. New York: Plenum; 1978. p. 57.

[22] McGarrity GJ, Sharma J, Vanamann V. Cell culture techniques. Am Soc Microbiol 1985a;51:170.

[23] McGarrity GJ, Kotani H. Cell culture mycoplasmas. In: Razin S, Barile MF, editors. The mycoplasmas;IV. Orlando, FL: Mycoplasma Pathogenicity Academic Press; 1985b.

[24] Ryan J. Understanding and managing cell culture contamination. Lowell, MA: Corning Incorporated, Life Sciences; 1994.

[25] Pauwels K, Herman P, Vaerenbergh BV, Thi CDD, Berghmans L, Waeterloos G, Bockstaele DV, Karoline Dorsch-Häsler K, Sneyers M. Animal cell cultures: risk assessment and bio-safety recommendations. Appl Biosaf 2007;12:26–38.

[26] Erickson GA, Bolin SR, Landgraf JG. Viral contamination of fetal bovine serum used for tis-sue culture: risks and concerns. Dev Biol Standard 1991;75:173–5.

[27] Zabal O, Kobrak AL, Lager IA, Schudel AA, Weber EL. Contamination of bovine fetal serum with bovine viral diarrhea virus. Rev Argent Microbiol 2000;32:27–32.

[28] Stoker AW, Hatier C, Bissell MJ. The embryonic environment strongly attenuates v-src onco-genes in mesenchymal and epithelial tissues, but not in endothelia. J Cell Biol 1990;111:217–28.

[29] Cunningham DA, Dos Santos Cruz GJ, Fernandez-Suarez XM, Whittam AJ, Herring C, Copeman L, Richards A, Langford GA. Activation of primary porcine endothelial cells induces release of porcine endogenous retroviruses. Transplantation 2004;77:1071–9.

[30] Anders M, Hansen R, Ding RX, Rauen KA, Bissell MJ, Korn WM. Distribution of 3D tissue integrity facilitates adenovirus infection by deregulating the Coxsackie virus and adenovirus receptor. Proc Natl Acad Sci 2003;100:1943–8.

[31] Vincent T, Pettersson RF, Crystal RG, LeopolD PL. Cytokine mediated downregulation of cox-sackievirus-adenovirus receptor in endothelial cells. J Virol 2004;78:8047–58.

[32] Linde AG. Safety instructions 12 (08.02), working with carbon dioxide CO_2. Germany: Gas and Engineering, Linde Gas Division. Available from: http://www.lindegas.hu/internet.lg.lg.hun/en/images/1270_15655.pdf?v=3.0.

[33] Cook Book. Fundamental techniques in cell culture laboratory handbook. 2nd ed. Cell types & culture characteristics, vol. 12. September 2010.

[34] Froehlich P. Noise pollution in the laboratory. Parker Hannifin Corporation; August 2013.

[35] Barkley WE. Safety considerations in the cell culture laboratory. Methods Enzym 1979;58:36–43.

[36] Caputo JL. Biosafety procedures in cell culture. J Tissue Cult Methods 1988;11:233–7.

[37] Grizzle WE, Polt S. Guidelines to avoid personnel contamination by infective agents in research laboratories that use human tissues. J Tissue Cult Methods 1988;11:191–9.

[38] McGarrity GJ. Spread and control of mycoplasma infection of cell culture. In Vitro 1976;12:643–8.

[39] Schmid I, Nicholson JKA, Giorgi JV, Janossy G, Kunkl A, Lopez PA, Perfetto S, Seamer LC, Dean PN. Biosafety guidelines for sorting of unfixed cells, special reports. Cytometry 1997;28:99–117.

[40] Lennartz K, Lu M, Flasshove M, Moritz T, Kirstein U. Improving the biosafety of cell sorting by adaptation of a cell sorting system to a biosafety cabinet. Cytometry 2005;66:119–27.

[41] Perfetto SP, Ambrozak DR, Roederer M, Koup RA. Viable infectious cell sorting in a BSL-3 facility. Methods Mol Biol 2004;263:419–24.

[42] CFR – Code of federal regulations, title 29: labor, part 1910—occupational safety and health standards. Subpart Z—Toxic Hazard Subst §1910.1096 Ioniz Radiation December 2014 (29-CFR 1910.1096).

[43] CFR – Code of Federal Regulations, NRC Regulation (CFR 10), § 35.12 Application for license, amendment, or renewal (10 CFR 35.12).

[44] CFR – Code of Federal Regulations, NRC Regulation (CFR 10), § 31.11 General license for use of byproduct material for certain in vitro clinical or laboratory testing (10 CFR 31.11).

[45] OSHA, laboratory safety guidance. Occupational Safety and Health Administration U.S. Department of Labor OSHA 3404-11; 2011.

[46] CFR – Code of federal regulations, title 29: labor, part 1910—occupational safety and health standards. Subpart Z—Toxic Hazard Subst §1910.1450 Occup Exposure Hazard Chem Lab December 2014 (29-CFR 1910.1450).

[47] CFR – Code of federal regulations, title 29: labor, part 1910—occupational safety and health standards. Subpart Z—Toxic Hazard Subst §1910.1200, Hazard Commun December 2014 (29-CFR 1910.1200).

[48] Occupational exposure to blood borne pathogens. Final rule. Standard interpretations: applicability of 1910.1030 to established human cell lines. 1991. 29 C.F.R. Sect. 1910.1030.

FURTHER READING

[1] Stacey G. Cell lines used in the manufacture of biological products. In: Spier R, editor. Encyclopedia of cell technology. New York: Wiley Interscienc; 2000. p. 79–83.

Chapter 11

In Vitro Methods for Predicting Ocular Irritation

Sanae Takeuchi[1], Seok Kwon[2]

[1]P&G Innovation Godo Kaisha, Kobe, Japan; [2]Procter & Gamble International Operations SA Singapore Branch, Singapore

INTRODUCTION

The eye is an organ that is exposed to external substances including hazardous chemicals. Possible adverse effects of chemicals on eyes can range from transient eye irritation to a significant loss of vision in a worst case scenario. The need to protect workers and consumers from adverse ocular effects associated with their exposure to chemicals makes evaluation of eye irritation potential, a critical step in the overall safety assessment process.

For over two decades, in vitro eye irritation tests have been developed for the purpose of classifying eye damage/eye irritation resulting from chemical exposure and several of which have been adopted by the Organization for Economic Co-operation and Development (OECD) as Test Guidelines (TG).

Historically, the Draize eye test [1] has, for many decades, been widely accepted as the in vivo test method to evaluate ocular effects induced by chemicals. The Draize eye test can evaluate specific ocular tissue effects on cornea, iris, and conjunctiva and the duration over which the tissue effect persists. In the Draize eye test, a test material (0.1 mL for liquid or 100 mg for solid) is instilled in the lower conjunctival sac of one eye of animals with the contralateral eye serving as an untreated control. Tissue effects on cornea (opacity), iris (inflammation), and conjunctiva (redness and swelling) are scored based on the defined grading system 1 h after the instillation followed by daily evaluation for 3 days and then on days 7, 14, and 21. To determine reversibility of the effects, the observation is normally conducted for 21 days. The evaluation should be terminated at a time when reversibility is confirmed before 21 days. The Draize eye test method is accepted by regulatory systems, such as the United Nations Globally Harmonized System of Classification and Labeling of Chemicals (UN GHS) classification in which ocular tissue scores and resolution of effects (persistence) are used to categorize chemicals (substances and mixtures) into UN GHS ocular

In Vitro Toxicology. http://dx.doi.org/10.1016/B978-0-12-804667-8.00011-0

irritation classes [2]. The Draize eye test was first adopted as an OECD TG (TG 405) in 1981 [3] with several revisions made thereafter and the latest revision of which includes the use of topical anesthetics, systemic analgesics, and humane endpoints in the procedures and measurements [4].

Although the Draize test has a very long use history, this method has been criticized in regard to its use of subjective evaluation criteria, physiological differences between test animals and humans, variability of scoring tissue effects, overprediction, and ethical concern regarding use of animals [5–9]. In addition, the recent legislative activity in several geographies, starting with the European Union animal testing ban in the seventh amendment to the Cosmetics Directive [10], added impetus to the development/adoption of alternative methods to replace the Draize eye test.

Evaluation of eye irritation using a testing strategy approach that combines in vitro test methods has been proposed for hazard identification and classification/labeling purposes by reducing the need for animals or completely replacing them [11]. The proposed "bottom-up" and "top-down" approach utilizes existing information on a chemical to estimate its eye irritation potential prior to progressing into further in vitro testing:

- the "bottom-up" approach, which starts with using in vitro test methods that can accurately identify chemicals not requiring classification for eye hazards according to UN GHS (UN GHS No Category);
- the "top-down" approach, which starts with using in vitro test methods that can accurately identify chemicals inducing serious eye damage/irreversible effects on the eye (UN GHS Category 1).

This chapter summarizes available eye irritation in vitro test methods, which have been validated and adopted as OECD TGs, and also reviews challenges with currently available in vitro test methods and ongoing developments in the area of in vitro test methods for the evaluation of ocular effects.

IN VITRO METHODS ADOPTED BY OECD

Isolated Chicken Eye Test

The isolated chicken eye (ICE) test method is an organotypic model that provides short-term maintenance of normal physiological and biochemical function of the chicken eye in vitro. The ICE protocol was developed by Prinsen and Koëter [12] based on the isolated rabbit eye test [13]. In the ICE test method, a test chemical is applied for 10s onto the cornea of eyes isolated from chickens that have been processed for human consumption. Three parameters are evaluated for up to 240 min to measure the extent of the eye effects following exposure of a test chemical to eyes. These are corneal swelling, corneal opacity, and fluorescein retention. The ICE test measures eye effects both quantitatively (corneal swelling) and also qualitatively (corneal opacity and fluorescein retention). Each endpoint measurement is categorized according to qualitative ICE classes (I–IV) based on predetermined scoring ranges, which are then combined together and used to assign a classification.

OECD initially adopted ICE test as an in vitro test method to identify chemicals inducing serious eye damage (UN GHS Category 1) in the "top-down" approach [14] and later also adopted it as a method to identify chemicals, which do not require eye irritation labeling (UN GHS No Category) in the "bottom-up" approach [15] (TG 483). The ICE test is applicable to solids (soluble or insoluble), liquid (aqueous or nonaqueous), emulsion, and gels for both the "top-down" and the "bottom-up" approach. Gases and aerosols have not been assessed in a validation study. Because in the "top-down" approach, high false positive rate for alcohols was reported in the OECD Streamlined Summary Document (SSD) [16], a user is expected to determine whether or not a potential overprediction for alcohols is acceptable or further testing can be performed in a weight-of-evidence (WoE) approach. Possible limitations for solids and surfactants are also identified in OECD 438 [15]. Availability of good quality eyes is one of the critical requirements for a testing laboratory to conduct the ICE test. The ICE test employs the qualitative measurement of corneal opacity and fluorescein retention, which could contribute to variation in the endpoint measurement. Thus, technical competence of the endpoint measurement is essential.

Histopathological examination of the corneas of eyes treated with a test material has been conducted as an additional endpoint to the ICE test for classification of nonextreme pH detergent and cleaning products and demonstrated to improve the predictive capability for UN GHS Category 1 of such formulations [17], thereby further elaborating the applicability domain [18].

Bovine Corneal Opacity and Permeability Test

The bovine corneal opacity and permeability (BCOP) assay is an organotypic eye irritation test method developed by Gautheron et al. [19]. It is a modification of an earlier ocular irritation assay, which uses isolated eyes of cattle that have been slaughtered for food consumption or other purposes [20]. The cornea is one of the main target tissues during accidental eye exposure to chemicals, and effects to the cornea can, in the worst case, result in visual impairment or loss. Therefore, it is a primary focus to assess corneal effects of a test chemical. Initially, corneal opacity was investigated because it is the endpoint graded for the cornea in in vivo ocular irritancy assays such as the Draize eye test. On the other hand, it is known that some substances, such as sodium lauryl sulfate and certain medium chain length alcohols, may affect the corneal epithelium and its barrier function without producing any significant corneal opacity. Tchao [21] proposed a method to quantify such effects on the epithelium by measuring corneal permeability. Gautheron and colleagues refined the BCOP assay to measure both opacity and permeability to predict ocular irritancy of a test chemical [19]. The IVIS (In Vitro Irritation Score) value, which is calculated from the opacity and permeability in a set formula, is used in the prediction model. Chemicals with IVIS >55 are identified as UN GHS Category 1, whereas chemicals with IVIS ≤3 are identified as UN GHS no category.

OECD initially adopted the BCOP test as a test method to identify UN GHS Category 1 chemicals in the "top-down" approach [22] and later also adopted as a method to identify GHS No Category chemicals in the "bottom-up" approach [23] (TG 437). The BCOP test can be applicable to any forms (solid/liquid) and/or class of chemicals for both the "top-down" and the "bottom-up" approach, whereas different protocols (exposure duration, postincubation period, etc.) may be used depending upon the forms (liquid/solid) of a test chemical. Because in the "top-down" approach high false positive rate for alcohols and ketones was reported in the OECD SSD [24], a user is expected to determine whether a potential overprediction for the group of chemicals is acceptable or further testing can be performed in a WoE approach. Possible limitations for solids are also identified in OECD 437 [23]. The availability of good quality eyes is one of the critical requirements for a laboratory to conduct the BCOP test. The quantitative measurement of endpoints is advantageous for the BCOP test.

Fluorescein Leakage Assay

The fluorescein leakage (FL) assay is a cytotoxicity and cell-function-based in vitro assay developed by Tchao [21]. The FL assay is performed on a confluent monolayer of MDCK (Madin–Darby canine kidney) cells on the insert membrane. The monolayer of MDCK cells forms intercellular tight junctions and desmosomal junctions, which are similar to those found on the apical side of conjunctival and corneal epithelia [25]. In in vivo systems, tight junctions and desmosomal junctions prevent chemicals from penetrating the corneal epithelium. Ocular irritants may damage the tight junctions and desmosomal junctions by inducing loss of transepithelial impermeability. In the FL assay, ocular irritancy of a chemical is evaluated by its ability to induce damage to an impermeable confluent epithelial monolayer by measuring an increase in permeability of sodium fluorescein through the monolayer. A test chemical in a solution at fixed concentration is applied to the confluent monolayer cells on the apical side of the insert for 1 min. The test chemical is then removed and sodium fluorescein dye is added to the apical side of the monolayer for 30 min. The amount of the sodium fluorescein dye that passes through the monolayer and the insert membrane is determined by measuring spectrofluorometrically the sodium fluorescein concentration in the well. A concentration of a test chemical causing 20% fluorescein leakage relative to blank (untreated) and maximum leakage (insert without cells) controls, i.e., FL_{20} is calculated and used in the prediction model for the identification of chemicals inducing serious eye damage/irreversible effects on eyes ($FL_{20} \leq 100 \, mg/mL$).

The FL assay was adopted by OECD as an alternative method to identify UN GHS Category 1 chemicals in the "top-down" approach [26] (TG 460).

A limitation for FL assay excludes test materials that have high acidity or alkalinity, cell fixative properties, highly volatile, and reactivity with medium contents from the applicability domain. Because a chemical is tested in a solution, the solubility of a chemical would be also a limiting factor for the applicability of the FL assay.

Short Time Exposure Test

The short time exposure (STE) test is a cytotoxicity-based in vitro assay to identify ocular irritants developed by Takahashi et al. [27]. The STE test method assesses cytotoxicity in a rabbit corneal epithelial cell line (SIRC cells) when 5% and 0.05% concentrations of a test chemical are applied for 5 min to a monolayer of SIRC cells cultured on 96-well microplate, by measuring relative viability of cells using the enzymatic conversion of the vital dye, 3-(4,5-dimethylthiazol-2-yl)-2,5-diphenyltetrazolium bromide (MTT) (MTT assay). When the cell viability is ≤70% at both 5% and 0.5% concentrations, a test chemical is classified as UN GHS Category 1. When cell viability is >70% at both concentrations, a test chemical is classified as UN GHS No Category.

The STE test method was adopted by OECD [28] as an alternative method used in both the "top-down" and the "bottom-up" approach (TG 492). In general, cytotoxicity tests using cultured cells such as the STE test are simple with rapid procedures at a low cost. A laboratory is required to have equipment for cell culture to conduct the STE test method. Water solubility is not a limiting factor for the applicability domain as the STE test method uses mineral oil as one of the vehicle options for poor water-soluble chemicals such as toluene, octanol, and hexanol. For the "bottom-up" approach, (1) highly volatile substances with a vapor pressure over 6 kPa and (2) solid chemicals (substances and mixtures) other than surfactants and mixtures composed only of surfactants are excluded from the applicability domain.

Reconstructed Human Cornea-Like Epithelium Test Method

The reconstructed human cornea-like epithelium (RhCE) test method is a cytotoxicity-based in vitro assay, which is performed on a three-dimensional RhCE tissue model derived from human cells. Although several RhCE tissue constructs are available commercially, to date only one tissue model [29] has undergone validation and been adopted by OECD to be used in the "bottom-up" approach [30]. Validation and regulatory acceptance of "me-too" RhCE test methods are in progress and it is expected that more RhCE test methods will be validated and adopted by OECD in the near future [31,32].

In the RhCE test method, a test chemical is applied directly onto the epithelial surface of the RhCE tissue construct in a way that is similar to how the corneal epithelium in vivo would be exposed to a foreign material. Chemicals that do not decrease tissue viability below a defined threshold are considered as UN GHS No Category. RhCE test method is applicable to any forms (liquid/ solid) or groups of chemicals, except gases and aerosols. Although a laboratory requires equipment for cell culture to conduct the RhCE test, the RhCE tissue model needs to be purchased from the manufacturer when the RhCE test is conducted. The RhCE tissue model manufacturer has to implement a robust quality control system to sustain a continuous production of quality tissue models.

RECENT DEVELOPMENT AND LOOKING INTO FUTURE

In vitro methods adopted by OECD can identify UN GHS Category 1 and/or No Category chemicals in the "top-down" or "bottom-up" approach. For chemicals that are not identified as UN GHS Category 1 or no category in these tests, additional tests are required to determine the final eye irritation classification. There is currently no in vitro method to specifically identify UN GHS Category 2 chemicals adopted by OECD. The primary focus of the development of in vitro methods is to better predict eye irritation potential of a chemical in the "top-down" and "bottom-up" approach as well as to identify UN GHS Category 2 chemicals.

In addition to OECD adopted methods, several other in vitro methods have been proposed. The porcine corneal ocular reversibility assay (PorCORA) is an organotypic assay in which porcine corneas are maintained in an air-interface culture model for 21 days to determine reversibility of corneal effects by fluorescein staining [33]. Study results with 32 chemicals reported a high correlation in the recovery time between PorCORA and the Draize eye test, which was correlated to the predictive capacity of PorCORA to identify GHS Category 1 chemicals [34]. The ex vivo eye irritation test (EVEIT) is another organotypic assay using isolated rabbit corneas [35]. Spöler et al. [36] used the method to develop a prediction model for UN GHS classification with 37 chemicals. One to three different chemicals and one positive control were tested using four application spots on one cornea culture. Untreated area of the cornea served as a negative control. Qualitative parameters including macroscopic opacity, fluorescein staining, and depth of damage using optical coherence tomography (OCT) are measured at $t \leq 1$ and $64\,h \leq t \leq 72\,h$ by which both acute corneal effects and persistence of the effects can be assessed. This study reported that EVEIT could be useful for either "top-down" or "bottom-up" approach with further simplification on the prediction model and with more data set. Tandon et al. [37] studied the predictive capacity of a reconstructed human three-dimensional hemi-cornea model, which comprises immortalized human corneal epithelial cells and keratocytes embedded in a collagen stroma. Two parameters were evaluated: (1) cell viability by MTT assay for both epithelial and stroma and (2) depth of injury of a section of the model using a microscope. The results indicated that there was a concordance in UN GHS classification (Category 1, 2, or no category) in 13 out of 14 chemicals using these two parameters. The cytosensor microphysiometer (CM) test is a cytotoxicity and cell-function based in vitro assay performed on a monolayer of adherent mouse L929 fibroblasts culture in the chamber of the

CM system [38]. The cells are exposed to a series of increasing concentration of a test chemical, and the metabolic activity of the cell is measured as the rate of acidification of the cell culture medium (production of acid metabolites) compared to the basal metabolic state. The concentration of the test chemical that reduces the acidification rate (i.e., metabolic rate) to 50%, i.e., MRD_{50}, is calculated and compared to cut-off values to determine eye irritation classification. The CM test is proposed to identify GHS category 1 and no category chemicals in the "top-down" and the "bottom-up" approaches, respectively. As a test chemical needs to be soluble in the cell culture medium, the applicability domain is limited to water-soluble test materials. For the "bottom-up" approach, the applicability domain is further limited to only water-soluble surfactants and surfactant-containing mixtures. A draft OECD TG for the CM test was issued in 2012 for public comment and remains in draft form [39].

Recent developments in the area of in vitro methods include noncell-based methods. For example, the Ocular Irritation assay utilizes an ordered macromolecular matrix (a mixture of proteins, glycoproteins, carbohydrates, lipids, and low molecular weight components) mimicking the structure of the transparent cornea to determine ocular irritation potency of chemicals by measuring a change in turbidity [40]. A validation study with 88 chemicals reported that the method could identify GHS Category 1 and no category in "top-down" and "bottom-up" approach, respectively, which is similar to in vitro methods adopted by OECD.

Review articles on the status of in vitro eye irritation test methods are available where more in-depth information is provided on the different methods including ones adopted by OECD as well as new methods reviewed in this chapter [41,42].

Any newly developed methods need to undergo validation before their regulatory acceptance. In the validation process, relevance (accuracy including sensitivity, specificity, positive and negative predictivity, false positive and false negative rates) and reliability (transferability and reproducibility) are assessed [43]. In the case of a new eye irritation test method, the Draize eye test would be used as an in vivo reference to assess the accuracy of a new method. Cosmetic Europe has compiled a database of the Draize data (Draize eye test Reference Database, DRD) from external lists that were used to support previous validation activities. Overall, the DRD contains Draize data on 634 individual chemicals from 681 independent in vivo studies and analyzed for the main ocular effects driving their eye irritation classification in the UN GHS classification system [9]. It was revealed that the most important drivers for UN GHS Category 1 classification are a mean value of corneal opacity for days 1–3 ($\geq$3) and persistence of corneal opacity on day 21, and those for Category 2 are a mean value of corneal opacity for days 1–3 ($\geq$1) and a mean value of conjunctival redness for days 1–3 ($\geq$2). Furthermore, it was considered that all effects should be present in $\geq$60% of the animals to drive a classification. Based on the in-depth analysis in this work, key criteria were identified, which should be considered when reference chemicals are selected in evaluation and validation of a new in vitro eye irritation test method. Combined with these considerations, the DRD is a useful tool for selecting reference chemicals for this purpose.

Evaluation of eye irritation potential is a critical step in the overall safety assessment process for chemicals. Not all chemicals have relevant in vivo or in vitro data for their eye irritation classification. UN GHS [2] describes a tiered approach on how to organize existing information and to make a WoE decision for eye irritation classification. OECD TG 405 [4] recommends a sequential testing and evaluation strategy. The sequence starts first from performing a WoE analysis on the existing relevant information of a chemical including human/ animal/in vitro data on eye effects, skin corrosion/irritation effects, QSAR, pH to name a few of the key data sources. When the analysis concludes that the available data are insufficient to determine eye irritation classification, then sequential testing utilizing adopted in vitro test methods, either "bottom-up" or "top-down" approach, is undertaken. This avoids or minimizes unnecessary in vivo tests. Currently, OECD is developing a Guidance Document to formalize the framework as an Integrated Approach on Testing and Assessment (IATA) for the purpose of eye hazard classification [44]. The draft Guidance Document indicates that IATA groups the various individual information sources as "Modules" based on the type of information provided (nine modules, in total), which are grouped in three major parts within the IATA; Part 1: existing and nontesting data, Part 2: WoE analysis, Part 3: new testing. once the guidance document is finalized, it would provide a consistent and transparent guidance on how to integrate existing information, generate data using in vitro methods, several of which were reviewed, in this chapter, in order to make a final decision on eye irritation classification of chemicals.

CONCLUSION

Several in vitro eye irritation test methods have been developed to replace the historic in vivo test, the Draize eye test. They include organotypic, cytotoxicity, and cell-function based, RhCE, and noncell based test methods. Some of these in vitro test methods have been adopted by OECD as TGs for eye hazard classification under regulatory classification system UN GHS. However, they are nonanimal partial replacement methods. Further work on the development/optimization of in vitro test methods to reflect better predictive capacity, with limited or no restrictions of applicability domain, is required to maximize their use in testing strategies to enable prediction covering all classification categories.

REFERENCES

[1] Draize JH, Woodard G, Calvery HO. Methods for the study of irritation and toxicity of substances applied topically to the skin and mucous membranes. J Pharmacol Exp 1944;82:377–90.

[2] UN. Globally harmonized system of classification and labelling of chemicals (GHS). ST/SG/ AC.10/30. 5th revised ed. New York and Geneva: United Nations; 2013.

[3] OECD. Guideline for testing of chemicals (No. 405). Acute Eye Irritation/Corrosion. Paris: Organisation for Economic Co-operation and Development; 1981.

[4] OECD. Guideline for the testing of chemicals (No. 405). Acute Eye Irritation/Corrosion. Paris: Organisation for Economic Co-operation and Development; 2012.

[5] Curren RD, Harbell JW. In vitro alternatives for ocular irritation. Environ Health Perspect 1998;106(Suppl. 2):485–92.

[6] York M, Steiling W. A critical review of the assessment of eye irritation potential using the draize rabbit eye test. J Appl Toxicol 1998;18:233–40.

[7] Wilhelmus KR. The draize eye test. Surv Ophthalmol 2001;45:493–515.

[8] Adriaens E, Barroso J, Eskes C, Hoffmann S, McNamee P, Alépée N, et al. Retrospective analysis of the Draize test for serious eye damage/eye irritation: importance of understanding the in vitro endpoints under UN GHS/EU CLP for the development and evaluation of in vitro test methods. Arch Toxicol 2014;88:701–23.

[9] Barroso J, Pfannenbecker U, Adriaens E, Alépée N, Cluzel M, De Smedt A, et al. Cosmetic Europe compilation of historical serious eye damage/eye irritation in vivo data analysed by drivers of classification to support the selection of chemicals for development and evaluation of alternative methods/strategies: the Draize eye test Reference Database (DRD). Arch Toxicol 2016. http://dx.doi.org/10.1007/s00204-016-1679-x.

[10] EU. EC – directive 2003/15/EC of the European parliament and of the council of 27 February 2003 amending council directive 76/768/EEC on the approximation of the laws of the member states relating to cosmetic products. Official J Eur Union March 11, 2003;L66:26.

[11] Scott L, Eskes C, Hoffmann S, Adriaens E, Alépée N, Bufo M, et al. A proposed eye irritation testing strategy to reduce and replace in vivo studies using bottom-up and top-down approaches. Toxicol In Vitro 2010;24:1–9.

[12] Prinsen MK, Koëter HBWM. Justification of the enucleated eye test with eyes of slaughterhouse animals as an alternative to the Draize eye irritation test with rabbits. Food Chem Toxicol 1993;31:69–76.

[13] Burton ABG, York M, Lawrence RS. The in vitro assessment of severe eye irritants. Food ChemToxicol 1981;19:471–80.

[14] OECD. Guideline for the testing of chemicals (No. 438). Isolated chicken eye test method for identifying ocular corrosives and severe irritants. Paris: Organisation for Economic Co-operation and Development; 2009.

[15] OECD. Guideline for the testing of chemicals (No. 438). Isolated chicken eye test method for identifying i) chemicals inducing serious eye damage and ii) chemicals not requiring classification for eye irritation or serious eye damage. Paris: Organisation for Economic Co-operation and Development; 2013.

[16] OECD. Streamlined summary document supporting OECD test guideline 438 on the isolated chicken eye for eye irritation/corrosion. 2013. Env/jm/mono. 12/PART 1. Paris: Organisation for Economic Co-operation and Development; 2013.

[17] Cazelle E, Eskes C, Hermann M, Jones P, McNamee P, Prinsen M, et al. Suitability of histopathology as an additional endpoint to the Isolated Chicken Eye Test for classification of non-extreme pH detergent and cleaning products. Toxicol In Vitro 2014;28:657–66.

[18] Cazelle E, Eskes C, Hermann M, Jones P, McNamee P, Prinsen M, et al. Suitability of the isolated chicken eye test for extreme pH detergent and cleaning products. Toxicol In Vitro 2015;29:609–16.

[19] Gautheron P, Dukic M, Alix D, Sina JF. Bovine corneal opacity and permeability test: an in vitro assay of ocular irritancy. Fundam Appl Toxicol 1992;18:442–9.

[20] Muir CK. Opacity of the bovine cornea in vitro induced by surfactants and industrial chemicals compared with ocular irritancy in vivo. Toxicol Lett 1985;24:157–62.

[21] Tchao R. Trans-epithelial permeability of fluorescein in vitro as an assay to determine eye irritants. In: Goldberg AM, editor. Alternative methods in toxicology. Progress in in vitro toxicology, vol. 6. New York: Mary Ann Liebert, Inc.; 1988. p. 271–83.

[22] OECD. Guideline for the testing of chemicals (no. 437). Bovine corneal opacity and permeability test method for identifying ocular corrosives and Severe irritants. Paris: Organisation for Economic Co-operation and Development; 2009.

[23] OECD. Guideline for the testing of chemicals (no. 437). Bovine corneal opacity and permeability test method for identifying i) chemicals inducing serious eye damage and ii) chemicals not requiring classification for eye irritation or serious eye damage. Paris: Organisation for Economic Co-operation and Development; 2013.

[24] OECD. Streamlined summary document supporting OECD guideline 437 on the bovine corneal opacity and permeability for eye irritation/corrosion. Env/jm/mono(2013)13. Paris: Organisation for Economic Co-operation and Development; 2013.

[25] INVITTOX protocol no. 71/DB-ALM protocol no. 71. The fluorescein leakage test. Available from: https://eurl-ecvam.jrc.ec.europa.eu/validation-regulatory-acceptance/topical-toxicity/eye-irritation.

[26] OECD. Guideline for the testing of chemicals (No. 460). Fluorescein leakage test method for identifying ocular corrosives and severe irritants. Paris: Organisation for Economic Co-operation and Development; 2012.

[27] Takahashi Y, Koike M, Honda H, Ito Y, Sakaguchi H, Suzuki H, et al. Development of the short time exposure (STE) test: an in vitro eye irritation test using SIRC cells. Toxicol In Vitro 2008;22:760–70.

[28] OECD. Guideline for the testing of chemicals (no. 491). Short time exposure in vitro test method for identifying i) chemicals inducing serious eye damage and ii) chemicals not requiring classification for eye irritation or serious eye damage. Paris: Organisation for Economic Co-Operation and Development; 2015.

[29] Kaluzhny Y, Kandárová H, Hayden P, Kubilus J, d'Argembeau-Thornton L, Klausner M. Development of the EpiOcular™ eye irritation test for hazard identification and labelling of EyeIrritating chemicals in response to the requirements of the EU cosmetics directive and REACH legislaion. ATLA 2011;39:339–64.

[30] OECD. Guideline for the testing of chemicals (no. 492). Reconstructed human cornea-like epithelium (RhCE) test method for identifying chemicals not requiring classification and labelling for eye irritation or serious eye damage. Paris: Organisation for Economic Co-operation and Development; 2015.

[31] Alépée N, Adriaens E, Grandidier MH, Meloni M, Nardelli L, Vinall CJ, et al. Multi-laboratory evaluation of SkinEthic HCE test method for testing serious eye damage/eye irritation using solid chemicals and overall performance of the test method with regard to solid and liquid chemicals testing. Toxicol In Vitro 2016;34:55–70.

[32] Katoh M, Hamajima F, Ogasawara T, Hata K. Establishment of a new in vitro test method for evaluation of eye irritancy using a reconstructed human corneal epithelial model, LabCyte CORNEA-MODEL. Toxicol In Vitro 2013;27:2184–92.

[33] Piehl M, Gilotti A, Donovan A, DeGeorge G, Cerven D. Novel cultured porcine corneal irritancy assay with reversibility endpoint. Toxicol In Vitro 2010;24:231–9.

[34] Piehl M, Carathers M, Soda R, Cerven D, DeGeorge G. Porcine corneal ocular reversibility assay (PorCORA) predicts ocular damage and recovery for global regulatory agency hazard categories. Toxicol In Vitro 2011;25:1912–8.

[35] Frentz M, Goss M, Reim M, Schrage NF. Repeated exposure to benzalkonium chloride in the ex vivo eye irritation test (EVEIT): observation of isolated corneal damage and healing. ATLA 2008;36:25–32.

[36] Spöler F, Kray O, Kray S, Panfill C, Schrage NF. The ex vivo eye irritation test as an alternative test method for serious eye damage/eye irritation. ATLA 2015;43:163–79.

[37] Tandon R, Bartok M, Zorn-Kruppa M, Brandner JM, Gabel D, Engelke M. Assessment of the eye irritation potential of chemicals: a comparison study between two test methods based on human 3D hemi-cornea models. Toxicol In Vitro 2015;30:561–8.

[38] INVITTOX protocol no. 130/DB-ALM protocol no. 130. The cytosensor microphysiometer (CM) toxicity test. Available from: https://eurl-ecvam.jrc.ec.europa.eu/validation-regulatory-acceptance/topical-toxicity/eye-irritation.

[39] OECD. Draft guideline. The cytosensor microphysiometer test method: an in vitro method for indentify ocular corrosive and severe irritant chemicals as well as chemicals no classified as ocular irritants. December 21, 2012. Available from: http://www.oecd.org/env/ehs/testing/DRAFT%20Cytosensor%20TG%20(V9)%2021%20Dec%2012_clean.pdf.

[40] Eskes C, Hoffmann S, Facchini D, Ulmer R, Wang A, Flego M, et al. Validation study on the ocular irritection® assay for eye irritation testing. Toxicol In Vitro 2014;28:1046–65.

[41] Wilson SL, Ahearne M, Hopkinson A. An overview of current techniques for ocular toxicity testing. Toxicology 2015;327:32–46.

[42] Lotz C, Schmid FF, Rossi A, Kurdyn S, Kampik D, De Wever B, et al. Alternative methods for the replacement of eye irritation testing. ALTEX 2015;33(1):55–67.

[43] OECD. No. 34, guidance document on the validation and International acceptance of new or updated test methods for hazard assessment. Env/jm/mono(2005)14. Paris: Organisation for Economic Co-operation and Development; 2005.

[44] OECD. Draft guidance document on and integrated approach on testing and assessment (IATA) for serious eye damage and eye irritation. Draft version 1, version 16.12.2015. Available from: https://www.oecd.org/env/ehs/testing/IATA_Eye_Hazard_Draft_v16Dec15.pdf.

Index

'*Note*: Page numbers followed by "f" indicate figures and "t" indicate tables.'

A

Acetoxymethyl diacetylester of calcein
cytotoxicity assay, 13–14
Additive effect, 29–30
Adenosine triphosphate (ATP), 120–123
Aldehydes, 187–188
Alzheimer's disease (AD), 142, 147
Ames test, 28, 51–52, 52f, 77
Amyloid beta, 147
Antagonistic effect, 29–30
Apoptosis detection assay, 49, 50f
Autophagy, 145–146

B

Bacterial reverse mutation, 70
Bacterial reverse mutation test (TG 471),
70–71
precautions, 71
principle, 70–71
procedure, 71
Basal toxicity, 8–9
BCOP. *See* Bovine corneal opacity and
permeability (BCOP)
Biosafety laboratory (BSL-2), 189
Bisphenol A (BPA), 108
Bovine corneal opacity and permeability
(BCOP), 201–202
Bovine viral diarrhea virus (BVDV), 184
Bruce Ames, 51–52

C

Caenorhabditis elegans, 141–142
Carbon dioxide, 186–187
Carcinogenicity assessment, 161–163
chromosomal aberration test, 162
gene mutation test, 161–162
genotoxicity assessment, 161–162
nongenotoxic carcinogenicity endpoints, 162
receptor-mediated endocrine modifiers, 162
transformation assays, 162–163
Cardiotoxicity in vitro, 13

Cell cultures, 6–7, 24–25
altering culture conditions, 186
biosafety and risk management, 178–179
containment measures, 191, 191t
culture media, 184–185
ergonomic hazards, 188
factors, 180–181
fetal bovine serum (FBS), 185
growth rate, 23
ionizing radiations, 188
manipulations transmission, 185
morphology, 23
mycoplasmas, 183–184
noise hazards, 188–189
nonionizing radiations, 188
optimization, 22–23
origin, 180–181
OSHA standards
Blood Borne Pathogens Standard, 192
Control of Hazardous Energy
Standard, 193
Eye and Face Protection Standard, 193
Hand Protection Standard, 193
Hazard Communication Standards, 192
Laboratory Standards, 192
Personal Protective Equipment (PPE)
Standard, 192–193
Respiratory Protection Standard, 193
outside biosafety cabinets, 185–186
physical hazards, 188–189
plating efficiency, 23
recommended practices, 189–190
risk, 179–180
classification, 179
CO_2 cylinders, 186–187
cryobanks/liquid nitrogen, 186
toxic compounds, 187–188
source of cells, 180–181
state and federal regulations, 191–192
tissue/type, 180–181
viral contaminanation, 181–183
origin, 183
problems, 182–183

Cell suspension, 71
Cerebral organoids, 11–12
Chaperone-mediated autophagy (CMA),
    145–146
Chemical Hygiene Plan (CHP), 192
Chemical-induced perturbation, 44
Chemoinformatics approaches, 89–92
Chorioallantoic membrane (CAM), 28
CHP. *See* Chemical Hygiene Plan (CHP)
Chromosomal aberration assay, 47–48
Comet assay, 47
Comparative molecular field analysis
    (CoMFA), 88–89
Comprehensive transcriptional profiling, 108
Connectivity mapping (cMAP), 113
Culture media, 184–185
Cytotoxicity assessment, 158–161
    acute system toxicity, 158–159
    benefits, 161
    conventional and evolving techniques,
        159–161
    endpoints, 159
    microarray, 160
    proteomics studies, 160
    RT-PCR, 160
Cytotoxic drugs, 190

D

Data interpretation, 31–32
Density functional theory (DFT), 89–92
Descriptive statistics, 31–32
3-(4, 5-dimethylthiazol-2-yl)-2,5-diphenyltet-
    razolium bromide (MTT) assay, 45–46
DNA content analysis, 48–49
Draize eye test, 199–202, 204–206
Drug induced liver injury (DILI), 120–123

E

Early growth response 1 (EGR1), 109–111
ECVAM. *See* European Centre for the
    Validation of Alternative Methods
    (ECVAM)
Embryonic stem (ES) cells, 123
Environmental Protection Agency's (EPA's),
    39, 106
Enzyme-linked immune sorbent assay, 44
Ergonomic hazards, 188
Escherichia coli, 70
17β-estradiol, 104–105
Estrogenic activity
    bisphenol A (BPA), 108

comprehensive transcriptional
    profiling, 108
connectivity mapping (cMAP), 113
early growth response 1 (EGR1), 109–111
Environmental Protection Agency's (EPA's),
    106
17β-estradiol, 104–105
estrogen receptor (ER), 105
17 α-ethynyl estradiol (EE), 108, 110f
exogenous chemicals, 104
G protein-coupled estrogen receptor
    (GPER), 105
high-throughput screening (HTS), 106
hydroxysteroid 11-beta dehydrogenase 2
    (HSD11B2), 109–111
Kruppel-like factor 4 (KLF4), 109–111
mode of action, 112–113
N-myc downregulated gene family
    (NDRG), 109–111
toxicology in the 21st century (Tox21),
    106–107
transcriptional profiling, 107–112
transforming growth factor alpha (TGFA1),
    109
transforming growth factor beta (TGFβ), 105
uterine transcriptional response, 108–109
in vitro to in vivo extrapolation, 113–114
Estrogen receptor (ER), 105
17 α-ethynyl estradiol (EE), 108, 110f
European Centre for the Validation of
    Alternative Methods (ECVAM), 3–4
Exogenous chemicals, 104
Explant culture, 23
Extrapolation, 130

F

Fetal bovine serum (FBS), 185
Fluorescein leakage (FL), 202–203
Forward gene mutation assay, 50–51
Fund for the Replacement of Animals in
    Medical Experiments (FRAME), 32

G

Gamma-H2AX assay, 48
Gene mutation test
    bacteria, 161
    mammalian cells, 161–162
Genotoxicity
    definition, 61–62
    future prospective, 78
    importance, 62–65

limitations, 76–77
  false negative results, 77
  false positive results, 77
  faulty metabolic activator system, 76–77
new chemical entity (NCE), 62–65
Organization for Economic Cooperation and
  Development (OECD), 62–65
regulatory guidelines, 62–65, 66t
rodent hematopoietic cells, 69
standard test battery, 67–69
test for, 65–67
thymidine kinase (TK), 65f
in vitro tests, 69–70
  bacterial reverse mutation test (TG 471),
    63f, 70–71
  controls, 70
  exposure concentrations, 70
  hypoxanthine-guanine phosphoribosyl
    transferase (HPRT), 72–73
  media and culture conditions, 70
  metabolic activation, 69
  solvent/vehicle, 70
  test substance/preparation, 70
  TK gene, 74–76
  in vitro mammalian cell gene mutation
    tests, 64f, 72–76
  in vitro mammalian cell micronucleus test
    (TG 487), 64f, 74
  in vitro mammalian chromosome
    aberration test (TG 473), 63f, 71–72
  xanthine-guanine phosphoribosyl
    transferase transgene (gpt) (XPRT),
    72–73
Global reactivity descriptors (GRD), 89–92
GLPs. *See* Good Laboratory Practices (GLPs)
Good Laboratory Practices (GLPs), 166–167,
  167f
G protein-coupled estrogen receptor (GPER),
  105
Growth rate, 23

H

Hepatoxicity in vitro, 10–11, 120–123,
  164–165
Heptox, 131
HGPRT gene mutation assay, 50–51
High-throughput screening (HTS), 106
Histone deacetylase 6 (HDAC6), 145–146
Hsp70, 139–141, 140f
Human kidney-2 (HK2), 128
Human toxicity
  advantages, 123–128

defined, 119–120
disadvantages, 123–128, 129t
drug induced liver injury (DILI),
  120–123
embryonic stem (ES) cells, 123
extrapolation, 130
hepatotoxicity, 120–123
induced pluripotent stem cells
  (iPSC), 123
organ level toxicity, 120, 121t–122t
predict toxicity, 130–131
system level toxicity, 120, 121t–122t
in vitro assay limitation, 128–130
in vitro systems
  liver toxicity assessment, 123–128
  nephrotoxicity assessment, 128
Huntingtin protein, 148–149
Huntingtin (Htt) protein, 141
Huntington disease (HD), 141, 148–149
Hydroxysteroid 11-beta dehydrogenase 2
  (HSD11B2), 109–111

I

ICE test method. *See* Isolated chicken eye
  (ICE) test method
Induced pluripotent stem cells (iPSC), 123
Inducer effect, 76
Inferential statistics, 31–32
In silico toxicology, 85–87
  predictive toxicology
    databases and web tools, 92–98, 93t–97t
    descriptors, 89–92, 90t–91t
    QSAR, 87–89
Interagency Coordinating Committee
  on the Validation of Alternative
  Method, 39
International Workshop on Genotoxicity Test
  Procedure, 1999, 77
In vitro mammalian cell gene mutation tests,
  72–76
  precautions, 73
  principle, 72–73, 75
  procedure, 73, 76
In vitro mammalian cell micronucleus test
  (TG 487)
  precaution, 74
  principle, 74
  procedure, 74
In vitro mammalian chromosome aberration
  test (TG 473), 71–72
  precautions, 72
  procedure, 72

In vitro models
  cell cultures, 24–25
  development of biomarkers, 33–34
  explant culture, 23
  Omics approach, 34
  organotypic culture, 24
  three dimensional culture systems, 25
  toxicants screening assays, 25–29, 26t–27t
    Alamar blue, 29
    Ames test, 28
    BCOP test, 29
    corrosivity assays, 28
    dose-response relationship, 29–31, 30f
    HET-CAM test, 28
    3T3 neutral red uptake test, 28–29
    tetrazolium salt,
      3-(4,5-dimethyl-thiazol-2-yl)-2,5-di-
        phenyltetrazolium bromide assay, 29
In vitro models for toxicity assessment
  bioinformatics and computational
    toxicology, 54
  cellular and functional responses, 52–53
  challenges, 54–55
  cytotoxicity assessment, 45–47, 45t
    3-(4, 5-dimethylthiazol-2-yl)-2,5-diphe-
      nyltetrazolium bromide (MTT) assay,
      45–46
    lactate dehydrogenase (LDH) assay, 46
    neutral red dye uptake assay, 46
    propidium iodide uptake assay, 46
    trypan blue exclusion assay, 47
  future perspective, 55–56
  genotoxicity assessment, 47–49
    apoptosis detection assay, 49, 50f
    cell cycle, 48–49
    chromosomal aberration assay, 47–48
    DNA content analysis, 48–49
    gamma-H2AX assay, 48
    micronucleus (MN) assay, 47
    single-cell gel electrophoresis assay, 47
    sister chromatid exchange (SCE), 48
  integrated testing strategies (ITS), 54, 55f
  models
    cytotoxicity studies, 42, 43t
    immunotoxicity, 44
    liver toxicity, 42
    lung toxicity, 43–44
    neurotoxicity, 44
    specific toxicity, 42–44
  mutagenicity
    Ames test, 51–52, 52f
    forward gene mutation assay, 50–51
    HGPRT gene mutation assay, 50–51

    mouse lymphoma L5178Y, 51
    nonmammalian model, 51–52
  needs, 40–41
  Omics approach, 53
  protein/gene expression, 52–53
  toxicokinetic study, 52
  validation, 41–42
    criteria, 41
    regulatory acceptance criteria,
      41–42
In vitro toxicology
  basal toxicity, 8–9
  cell culture, 6–7
  cells to 3D organoids, 10–14
    cardiotoxicity in vitro, 13
    cerebral organoids, 11–12
    hepatoxicity in vitro, 10–11
    nanotoxicology in vitro, 13–14
    nephrotoxicity in vitro, 12–13
    neurotoxicity in vitro, 11–12
  cytotoxicity, 8–9
    acetoxymethyl diacetylester of calcein
      cytotoxicity assay, 9
    adenosine diphosphate ratio assay, 9
    adenosine triphosphate (ATP), 9
    adenosine triphosphate ratio assay, 9
    lactate dehydrogenase assay, 9
    neutral red uptake assay, 9
    tetrazolium bromide salt assay, 9
  definition, 1–2
  explant culture, 6
  genotoxicity, 10
  immortalized cell lines, 7
  induced pluripotent stem cells, 8
  organ culture, 6
  organoid culture, 8
  organotypic cultures, 7
  organ-specific endpoints, 8–10
  policy makers, 2–4
  prevalidation, 3
  primary cell culture, 7
  projects and developments, 4–5
  regulatory control, 2–4
  research and development, 3
  stem cells, 8
  test methods, 32–33
  types, 6
  validation, 3
  in vitro alternatives and methods, 5–6
  in vivo, 2
Ionizing radiations, 188
Isolated chicken eye (ICE) test method,
  200–201

K

Karyotyping, 22
Kruppel-like factor 4 (KLF4), 109–111

L

Lactate dehydrogenase (LDH) assay, 46
Lesch-Nyhan syndrome, 72
Liver toxicity assessment, 123–128
Local reactivity descriptors (LRD), 89–92
Long-term culturing/subculturing, 23
Lysosomal proteolysis, 145–146

M

mEST. *See* Mouse ES cell test (mEST)
Microbial contamination, 22
Micronucleus (MN) assay, 47
Molecular chaperones, 138–143, 140f
Morphology, 23
Mouse ES cell test (mEST), 163–164
Mouse lymphoma L5178Y, 51
Mycoplasma-infected host cells, 184
Mycoplasma pneumoniae, 183–184
Mycoplasmas, 183–184

N

Nanotoxicology in vitro, 13–14
Nephrotoxicity in vitro, 12–13, 128
Neurotoxicity in vitro, 11–12
Neutral red dye uptake assay, 46
New chemical entity (NCE), 62–65
Nicotinamide adenine dinucleotide (NAD), 46
N-myc downregulated gene family (NDRG),
    109–111
Noise hazards, 188–189
Nonionizing radiations, 188
Nonmammalian model, 51–52
Nucleotide exchange factor (NEF), 139–140

O

Occupational Safety and Health Act (OSH
    Act), 1970, 191
  Blood Borne Pathogens Standard, 192
  Control of Hazardous Energy Standard, 193
  Eye and Face Protection Standard, 193
  Hand Protection Standard, 193
  Hazard Communication Standards, 192
  Laboratory Standards, 192
  Personal Protective Equipment (PPE)
      Standard, 192–193
  Respiratory Protection Standard, 193

Ocular irritation
  bottom-up approach, 200
  Draize eye test, 199–202, 204–206
  evaluation, 200, 206
  Organization for Economic Co-operation
      and Development (OECD), 199
    bovine corneal opacity and permeability
        (BCOP), 201–202
    developments, 204–206
    fluorescein leakage (FL), 202–203
    future perspectives, 204–206
    isolated chicken eye (ICE) test method,
        200–201
    reconstructed human cornea-like
        epithelium (RhCE), 203–204
    short time exposure (STE), 203
    top-down approach, 200
Optimum test protocol, 40
Organization for Economic Co-operation and
    Development (OECD), 41–42
Organotypic culture, 24
Oxidative stress, 33–34

P

Parkinson's disease (PD), 139–140
Phenolic-based disinfectants, 187–188
Photosensitization, 28–29
Phototoxicity assessment, 164
Plating efficiency, 23
Potentiation effect, 29–30
Predictive toxicology
  databases and web tools, 92–98, 93t–97t
  descriptors, 89–92, 90t–91t
  QSAR, 87–89
Propidium iodide uptake assay, 46
Proteotoxicity
  Alzheimer's disease (AD), 142, 147
  amyloid beta, 147
  autophagy, 145–146
  Caenorhabditis elegans, 141–142
  cellular defense against, 138–139
  cellular degradation mechanism, 143
  chaperone-mediated autophagy (CMA),
      145–146
  degradation pathways, 146
  encephalopathies prion proteins, 149–150
  histone deacetylase 6 (HDAC6), 145–146
  Hsp70, 139–141, 140f
  Huntingtin protein, 148–149
  Huntingtin (Htt) protein, 141
  Huntington disease (HD), 141, 148–149
  lysosomal proteolysis, 145–146

Proteotoxicity (*Continued*)
  mechanisms, 137–138, 137f
  molecular chaperones, 138–143, 140f
  nucleotide exchange factor (NEF), 139–140
  Parkinson's disease (PD), 139–140
  protein folding and misfolding, 136–137
  proteotoxic proteins, 139–143
  resulted disorders, 136t
  Saccaromyces cerevisiae, 141–142
  sHsps, 142
  α-synuclein, 147–148
  ubiquitin proteasome system (UPS),
    143–145, 145f
Proteotoxic proteins, 139–143

Q

Quantitative structure activity relationships
  (QSAR), 85–87

R

Reconstructed human cornea-like epithelium
  (RhCE), 203–204
Regulatory acceptance, 170, 171t–173t
Relative survival (RS), 73
Reproductive toxicity assessment, 163–164
3 Rs principles
  reduction, 21
  refinement, 21
  replacement, 21

S

Saccaromyces cerevisiae, 141–142, 161
Safety pharmacology, 165
Safety/toxicity assessment
  animal testing, 170
  carcinogenicity assessment, 161–163
    chromosomal aberration test, 162
    gene mutation test, 161–162
    genotoxicity assessment, 161–162
    nongenotoxic carcinogenicity endpoints,
     162
    receptor-mediated endocrine modifiers,
     162
    transformation assays, 162–163
  cytotoxicity assessment, 158–161
    acute system toxicity, 158–159
    benefits, 161
    conventional and evolving techniques,
     159–161
    endpoints, 159

microarray, 160
  proteomics studies, 160
  RT-PCR, 160
  developmental toxicity assessment, 163–164
  Good Laboratory Practices (GLPs),
    166–167, 167f
  hepatotoxicity, 164–165
  mechanism, 157
  mouse ES cell test (mEST), 163–164
  phototoxicity assessment, 164
  regulatory acceptance, 170, 171t–173t
  reproductive toxicity assessment, 163–164
  safety pharmacology, 165
  tissue cross-reactivity assay, 165–166
  translational aspects, in vitro assays,
    170–174
    pitfalls, 170–174
  validation methods, 168–170, 168f
    accuracy, 170
    interlaboratory reproducibility, 170
    intralaboratory repeatability, 170
    intralaboratory reproducibility, 170
    relevance, 169–170
    reliability, 169
    repeatability, 169
    reproducibility, 169
    robust(ness), 169
    sensitivity, 169
    specificity, 169
  in vitro assays, 157–166
  whole embryo culture (WEC), 163–164
Salmonella typhimurium, 28, 70–71
Short time exposure (STE), 203
Single-cell gel electrophoresis assay, 47
Sister chromatid exchange (SCE), 48
Small nucleus, 74
Synergistic effect, 29–30
α-synuclein, 147–148
System level toxicity, 120, 121t–122t

T

6-thioguanine (6-TG), 73
6-thioguanosine monophosphate (TGMP), 73
Three-dimensional (3D) architecture
  mimicking organ behavior, 123
Three-dimensional correlation models
  (3D-QSAR), 88–89
Three dimensional culture systems, 25
Thymidine kinase (TK), 65–66
Tissue cross-reactivity assay, 165–166
Toxicity, 83–84

Toxicology in the 21st century (Tox21), 106–107

Transcriptional profiling, 107–112

Transforming growth factor alpha (TGFA1), 109

Transforming growth factor beta (TGFβ), 105

Trypan blue exclusion assay, 47

U

Ubiquitin proteasome system (UPS), 143–145, 145f

Uterine transcriptional response, 108–109

V

Viral contaminanation, 181–183

origin, 183

problems, 182–183

W

Whole embryo culture (WEC), 163–164

Printed in the United States
By Bookmasters